Chemistry for Beginners 2025

The Ultimate Handbook for Students to Effortlessly Understand and Master the Study of Elements and Compounds with Clarity and Enjoyment

Ethan D. Needleman

TABLE OF CONTENTS

INTRODUCTION

Have you ever felt overwhelmed by the complex jargon of chemistry? Do you struggle to grasp the fundamental concepts that seem crucial for understanding the subject? Are you frustrated by confusing diagrams and lengthy explanations that leave you more puzzled than enlightened? Do you find it challenging to relate abstract chemical principles to everyday life? Are you searching for a resource that makes learning chemistry not just easier, but also enjoyable?

If these questions resonate with you, then *Chemistry for Beginners 2025: The Ultimate Handbook for Students* is exactly what you need. This book is designed to address these very pain points, offering you a clear, engaging, and comprehensive introduction to the world of chemistry.

Chemistry can seem like an intricate maze of terms, theories, and equations. Yet, at its core, it's the study of the elements and compounds that make up everything around us. Our mission is to simplify this complexity, breaking down each concept into manageable, easy-to-understand segments. We aim to make chemistry not just a subject to be learned, but a fascinating adventure of discovery.

In the chapters ahead, you'll embark on a journey from the basic building blocks of matter to the intricate world of organic and inorganic chemistry. Whether you're navigating the structure of atoms, exploring the periodic table, or delving into chemical reactions, this handbook will guide you with clarity and purpose.

We've included practical tips for effective learning and real-life applications of chemistry to make your study both meaningful and enjoyable. From understanding the importance of chemistry in daily life to mastering laboratory techniques, this book provides a holistic approach that makes complex ideas accessible and engaging.

So, get ready to unravel the mysteries of chemistry with confidence and curiosity. Welcome to a learning experience that promises clarity, enjoyment, and mastery. Let's dive into the fascinating world of chemistry together!

Welcome to Chemistry

How to Use This Book

1. Structure and Organization

This book is divided into distinct chapters, each focusing on a specific aspect of chemistry. The chapters are arranged in a logical progression, starting from fundamental concepts and gradually moving towards more advanced topics. Here's a brief overview of how to navigate through the book:

Foundations of Chemistry: Begin with the basics to build a solid understanding of what chemistry is and why it matters.

The Basics of Matter: Learn about the different forms and properties of matter, setting the stage for deeper exploration.

Atomic Structure: Explore the building blocks of matter, including atoms and their components.

The Periodic Table: Understand how elements are organized and what trends and patterns reveal about their behavior.

Chemical Bonding: Delve into how atoms connect to form molecules and the nature of these bonds.

Chemical Reactions: Discover how substances transform and what factors influence these changes.

Each chapter builds upon the previous one, so it's helpful to read them in order. However, feel free to revisit sections as needed to reinforce your understanding or clarify concepts.

2. Key Concepts and Terms

At the beginning of each chapter, you'll find a list of key concepts and terms. These are essential for grasping the material covered in the chapter. Make sure to familiarize yourself with these terms as you read. They will help you understand the subject matter more deeply and will be useful for reference as you progress through the book.

3. Tips for Effective Learning

Active Reading: Engage with the material by taking notes, asking questions, and summarizing key points in your own words. This will help reinforce your understanding and retention of the information.

Practice Problems: Many chapters include practice problems or questions to test your knowledge. Work through these problems to apply what you've learned and gain confidence in your understanding.

Visual Aids: Use diagrams, charts, and illustrations provided in the book. Visual aids can make abstract concepts more concrete and easier to understand.

Hands-On Experiments: Where applicable, try to perform simple experiments or demonstrations. Experiencing chemistry firsthand can enhance your comprehension and make learning more enjoyable.

4. Chapter Summaries and Review

At the end of each chapter, you'll find a summary of the key points covered. These summaries are useful for reviewing the material and ensuring you've grasped the main concepts. Additionally, review questions at the end of each chapter help you assess your understanding and prepare for more advanced topics.

5. Glossary and Index

To assist with your studies, the book includes a glossary of key terms and an index. The glossary provides definitions of important terms used throughout the book, while the index allows you to quickly locate specific topics or concepts. Use these tools to enhance your study sessions and find information efficiently.

6. Interactive Elements

In this edition, we've incorporated interactive elements such as online resources or supplementary materials (if applicable). These can provide additional explanations, practice exercises, or visual demonstrations. Check the introduction or any supplementary sections for links or references to these resources.

7. Advanced Topics (Optional)

For those interested in exploring more advanced concepts, the final chapters cover optional topics like quantum chemistry and nanotechnology. These sections are designed for readers who want to delve deeper into specific areas of chemistry. Feel free to explore these topics if you're curious, but they're not required for a solid understanding of the basics.

By following these guidelines, you'll be well-equipped to tackle the challenges of learning chemistry. Approach each chapter with curiosity and an open mind, and remember that chemistry is a fascinating field with endless possibilities for discovery. Enjoy your journey into the world of chemistry!

Key Concepts and Terms

Understanding chemistry begins with mastering its foundational concepts and terminology. These key concepts and terms form the building blocks of the subject, providing the essential vocabulary and ideas you'll need to grasp more complex topics as you progress. Here's an extensive discussion on the key concepts and terms that are fundamental to the study of chemistry:

1. What is Chemistry?

Chemistry is the scientific discipline that studies the composition, structure, properties, and changes of matter. It explores how substances interact, combine, and change to form new substances. Chemistry is often referred to as the "central science" because it connects physical sciences with life sciences and applied sciences.

Matter: Anything that has mass and occupies space. Matter can exist in different states, including solids, liquids, and gases.

Substance: A form of matter that has a uniform and definite composition. Examples include elements and compounds.

Chemical: A substance that has a defined composition and properties.

2. States of Matter

States of Matter refer to the distinct forms that different phases of matter take on. The most common states are:

Solid: Has a fixed shape and volume. The particles are closely packed in a fixed arrangement.

Liquid: Has a definite volume but takes the shape of its container. The particles are close together but can move past each other.

Gas: Has neither a fixed shape nor a fixed volume. The particles are far apart and move freely.

3. Chemical Bonding

Chemical Bonding involves the attraction between atoms that allows the formation of chemical substances. The primary types of chemical bonds are:

Ionic Bond: Formed when one atom gives up one or more electrons to another atom, creating ions that attract each other.

Covalent Bond: Formed when two atoms share one or more pairs of electrons.

Metallic Bond: Involves the pooling of electrons among a lattice of metal atoms, allowing them to conduct electricity and heat.

Molecular Geometry and **VSEPR Theory** help predict the shapes of molecules based on electron pair repulsion.

4. The Periodic Table

The **Periodic Table** is a tabular arrangement of elements, organized by increasing atomic number and grouped by similar chemical properties.

Element: A pure substance made up of only one type of atom.

Atomic Number: The number of protons in the nucleus of an atom.

Periodic Trends: Patterns observed in the properties of elements, such as atomic radius, ionization energy, and electronegativity.

Groups (columns) and **Periods** (rows) categorize elements with similar properties.

5. Chemical Reactions

Chemical Reactions are processes where reactants transform into products, involving the breaking and forming of chemical bonds.

Reactants: Substances that undergo a chemical change.

Products: New substances formed as a result of the chemical reaction.

Balancing Chemical Equations: Ensuring that the number of atoms of each element is the same on both sides of the equation.

Catalysts: Substances that increase the rate of a reaction without being consumed.

6. Solutions and Concentrations

Solutions are homogeneous mixtures of two or more substances.

Solvent: The substance that dissolves the solute (usually the component in greater amount).

Solute: The substance that is dissolved in the solvent.

Concentration: The amount of solute present in a given quantity of solvent or solution, expressed in units like molarity.

Colligative Properties such as boiling point elevation and freezing point depression are affected by the concentration of solute particles.

7. Acids and Bases

Acids and **Bases** are classes of compounds with distinct properties and reactions.

Acid: A substance that donates protons (H^+ ions) in a solution. Common acids include hydrochloric acid (HCl) and sulfuric acid (H_2SO_4).

Base: A substance that accepts protons or donates hydroxide ions (OH^-). Examples include sodium hydroxide (NaOH) and ammonia (NH_3).

pH: A scale used to measure the acidity or basicity of a solution, ranging from 0 to 14.

Buffer Solutions help maintain a stable pH when small amounts of acids or bases are added.

8. Thermodynamics and Kinetics

Thermodynamics and **Kinetics** deal with energy changes and reaction rates, respectively.

Thermodynamics: Studies the heat and energy involved in chemical processes. Key concepts include **Enthalpy** (heat content) and **Entropy** (disorder).

Kinetics: Examines the speed of chemical reactions and the factors affecting reaction rates, such as temperature and concentration. **Activation Energy** is the energy required to initiate a reaction.

9. Organic and Inorganic Chemistry

Organic Chemistry focuses on carbon-containing compounds and their reactions.

Hydrocarbons: Compounds made of carbon and hydrogen, including alkanes, alkenes, and alkynes.

Functional Groups: Specific groups of atoms within molecules that determine their chemical reactivity, such as alcohols and carboxylic acids.

Inorganic Chemistry covers compounds not typically involving carbon, including metals, minerals, and coordination compounds.

10. Environmental and Industrial Chemistry

Environmental Chemistry studies the chemical processes occurring in the environment and their impact.

Pollution: The introduction of harmful substances into the environment.

Green Chemistry: Designing chemical processes that minimize environmental impact and reduce waste.

Industrial Chemistry involves the large-scale production of chemicals and materials, including pharmaceuticals and materials science.

Mastering these key concepts and terms provides a solid foundation for understanding and exploring the broader field of chemistry. By familiarizing yourself with this vocabulary and these principles, you'll be better equipped to tackle more complex topics and appreciate the intricate ways in which chemistry impacts our world.

Tips for Effective Learning

Learning chemistry can seem daunting at first, with its vast array of concepts, equations, and technical terminology. However, with the right approach, you can turn chemistry into an enjoyable and rewarding experience. This section is dedicated to providing you with effective strategies and tips for mastering chemistry, so you can understand and retain key concepts with ease. Whether you're just beginning your journey in chemistry or looking to enhance your study techniques, these tips will help you make the most of your learning experience.

1. Start with the Basics and Build a Strong Foundation

Chemistry, like mathematics, builds upon foundational principles. If you jump into advanced topics without fully understanding the basics, you're likely to become confused and frustrated. To avoid this, make sure you have a solid grasp of core concepts before moving on to more complex material. Some essential foundational topics include:

The structure of atoms and subatomic particles (protons, neutrons, and electrons)

Understanding elements, compounds, and mixtures

The periodic table and how elements are organized

Chemical bonding (ionic, covalent, and metallic bonds)

Once you're comfortable with these fundamental topics, you'll find it easier to grasp more advanced ideas such as reaction mechanisms, organic chemistry, or thermodynamics.

2. Practice Active Learning: Engage with the Material

Passive reading or skimming through a textbook is rarely enough to truly understand chemistry. **Active learning** means engaging with the material in a way that forces you to think critically and interact with the concepts. Here are some ways to practice active learning:

Ask Questions: As you read or study, ask yourself questions like "Why does this happen?" or "How does this concept apply to real-world situations?"

Summarize in Your Own Words: After learning a new concept, take a moment to summarize it in your own words. This forces you to process the information and helps reinforce your understanding.

Teach Someone Else: One of the best ways to solidify your understanding is to teach a concept to someone else. If you can explain it clearly, you know you've mastered it.

3. Practice Problem-Solving Regularly

Chemistry is not just about memorizing facts and equations; it's about applying knowledge to solve problems. Problem-solving is a key skill in chemistry, and regular practice will help you become proficient at it. Here's how to approach problem-solving effectively:

Work Through Practice Problems: Each chapter in this book will include practice problems. Make sure you complete these exercises to test your understanding.

Focus on Understanding, Not Just Memorization: When solving problems, don't just focus on getting the right answer. Understand the steps and reasoning behind each solution. This will help you apply similar methods to different problems in the future.

Break Down Complex Problems: If a problem seems too complex, break it down into smaller, manageable parts. Identify what is being asked, list the known information, and determine which concepts or equations are relevant.

4. Visualize Concepts with Diagrams and Models

Many concepts in chemistry, such as atomic structure, molecular geometry, and reaction mechanisms, are highly abstract and difficult to understand just through reading. Visual aids such as diagrams, models, and charts can make these concepts more tangible and easier to comprehend.

Use Diagrams: For topics like molecular structures, chemical reactions, or the periodic trends, drawing diagrams can help you visualize what's happening at the molecular or atomic level.

Build Models: If possible, use molecular model kits or online tools to build 3D representations of molecules. This can help you better understand how atoms bond and form specific shapes.

Use Concept Maps: Concept maps are diagrams that show how different ideas are connected. Creating concept maps can help you see the relationships between different chemistry topics.

5. Make Chemistry Relevant to Your Life

Chemistry can sometimes feel abstract and disconnected from everyday experiences, but it's deeply intertwined with the world around us. One of the best ways to stay motivated is to connect what you're learning with real-world applications. Here are a few ways to do this:

Look for Chemistry in Daily Life: Think about how chemistry affects your daily life. For example, consider how cleaning products work (acid-base chemistry), why food cooks the way it does (chemical reactions), or how medications work in your body (organic chemistry).

Relate Concepts to Technology: Chemistry plays a vital role in developing new technologies. Exploring topics like the chemistry behind batteries, materials used in smartphones, or the production of pharmaceuticals can give you a better appreciation of its practical value.

Perform Simple Experiments: If possible, try conducting safe, simple experiments at home or in a lab setting. Hands-on learning can make abstract concepts more concrete and engaging.

6. Stay Organized and Plan Your Study Sessions

Effective learning requires structure and organization, especially for a subject as broad as chemistry. To ensure you're making steady progress, it's important to have a plan in place for how you will study. Here are some tips for staying organized:

Create a Study Schedule: Break down your study material into manageable chunks and set a timeline for completing each section. For example, you might dedicate one week to mastering atomic structure and another to chemical bonding.

Use Checklists: Keep track of the topics you've covered and those you still need to work on. A checklist can help you stay on top of your learning goals.

Take Regular Breaks: Studying for long periods without breaks can lead to burnout. Use the Pomodoro Technique (25 minutes of study followed by a 5-minute break) or something similar to keep your focus sharp.

7. Review and Reinforce Regularly

Repetition is key to retaining information. Don't wait until the day before a test to review; make it a regular part of your study routine. Here's how to review effectively:

Regular Review Sessions: At the end of each week, go over the material you've learned to reinforce your understanding. Use chapter summaries and review questions to test your retention.

Use Flashcards: For memorizing important terms, chemical equations, or reaction mechanisms, flashcards can be an effective tool. You can use physical flashcards or apps like Anki or Quizlet.

Revisit Difficult Topics: If you encounter a concept that's difficult to understand, don't shy away from revisiting it. Chemistry often requires multiple exposures to a topic before it fully clicks.

8. Collaborate and Discuss with Others

Learning chemistry can sometimes feel isolating, but collaboration with classmates or study groups can enhance your learning experience. Discussing challenging concepts with others can provide new insights and help clarify misunderstandings.

Join a Study Group: Working with a study group allows you to exchange ideas, test each other's knowledge, and fill in gaps in understanding. Hearing others explain concepts can give you a new perspective.

Ask for Help When Needed: Don't hesitate to seek help from a teacher, tutor, or peer when you're struggling with a topic. A different explanation or approach can make all the difference.

9. Use Technology to Enhance Learning

Modern technology offers a wealth of resources that can supplement your learning in chemistry. Whether it's video tutorials, interactive simulations, or educational apps, these tools can make difficult concepts more accessible.

Watch Video Tutorials: Platforms like YouTube, Khan Academy, and Coursera offer free chemistry tutorials. Sometimes a visual demonstration or a different teaching style can make a complex topic easier to understand.

Use Simulations and Interactive Tools: Online tools like PhET simulations or ChemCollective allow you to experiment with virtual chemistry labs, visualize molecular structures, and explore chemical reactions in an interactive environment.

10. Stay Curious and Keep Exploring

Finally, remember that chemistry is a fascinating and ever-evolving field. Stay curious, ask questions, and seek out additional learning opportunities. The more you explore, the more you'll appreciate how chemistry shapes our world—from the smallest atom to the vast expanse of industrial processes and environmental systems.

With these tips, you'll be well-equipped to approach chemistry with confidence, curiosity, and a sense of enjoyment. Keep in mind that effective learning takes time and effort, but with patience and persistence, you'll gain a deep understanding of chemistry that will serve you well in both academics and life.

Chapter 1: Foundations of Chemistry

What is Chemistry?

Chemistry is a branch of science that studies matter—its composition, properties, structure, and the changes it undergoes during various processes. As one of the fundamental sciences, chemistry explores the interactions between atoms, molecules, and other forms of matter to explain the physical world. It is often referred to as the "central science" because it bridges other natural sciences like biology, physics, and environmental science, and it plays a key role in technological and industrial developments.

Understanding chemistry allows us to comprehend the building blocks of the universe and the principles governing the physical and chemical changes we observe daily. From the water we drink, the food we eat, and the air we breathe, to the products we use and the technologies that power our lives, chemistry is everywhere.

Let's dive deeper into the various dimensions of what chemistry encompasses:

1. The Study of Matter and Its Composition

At its core, chemistry is the study of matter. Matter is defined as anything that has mass and occupies space. It exists in different forms and can be as simple as a pure element like hydrogen or a complex mixture like the air we breathe.

Matter: Matter is anything that occupies space and has mass. Matter exists in different states, such as solids, liquids, gases, and plasmas, and is made up of atoms and molecules.

Atoms: Atoms are the basic building blocks of matter. Each element on the periodic table is made of a specific type of atom, characterized by a unique number of protons, neutrons, and electrons.

Molecules: Molecules are two or more atoms bonded together. They can be simple (like O_2) or complex (like proteins and DNA), and they form the basis of everything in the physical world.

Elements and Compounds: Elements are pure substances made of only one type of atom, such as oxygen (O) or carbon (C). Compounds, on the other hand, are substances made from two or more different types of atoms chemically bonded together, such as water (H_2O) or carbon dioxide (CO_2).

By understanding the composition of matter, chemists can predict how substances will behave and how they can be transformed through chemical reactions.

2. Properties of Matter

Chemistry investigates both the **physical** and **chemical properties** of substances.

Physical Properties: These are characteristics that can be observed or measured without changing the substance's composition. They include color, density, melting point, boiling point, and electrical conductivity. For instance, ice melts into water at 0°C, which is a physical property because the chemical identity of H_2O remains the same.

Chemical Properties: These refer to a substance's ability to undergo chemical changes that transform it into different substances. For example, iron's tendency to rust when exposed to oxygen and moisture is a chemical property because it results in the formation of a new substance, iron oxide (rust).

Chemistry helps us understand these properties and how they can be manipulated in both everyday and industrial applications.

3. The Structure of Matter

The study of chemistry involves understanding the internal structure of matter, particularly how atoms are arranged and how they interact.

Atomic Structure: Atoms consist of a central nucleus made of protons (positively charged particles) and neutrons (neutral particles), with electrons (negatively charged particles) orbiting around this nucleus. The way these subatomic particles are arranged influences the atom's properties and its behavior in chemical reactions.

Bonding: Atoms bond together to form molecules. The type of bond—whether it's ionic (transfer of electrons), covalent (sharing of electrons), or metallic (a sea of delocalized electrons)—affects the structure and properties of the resulting compounds.

4. Transformations of Matter

One of the most significant aspects of chemistry is understanding how matter changes. These changes can either be physical or chemical.

Physical Changes: Changes that affect the form of a substance but not its chemical composition. For example, melting ice into water or boiling water into steam involves physical changes because the substance remains H_2O throughout the process.

Chemical Changes: Changes that result in the formation of one or more new substances. For instance, when hydrogen reacts with oxygen to form water (H_2O), a chemical change occurs, and the properties of water are different from those of hydrogen and oxygen.

5. Energy and Matter

Another fundamental principle in chemistry is the role of energy in the behavior and transformation of matter.

Energy: Energy is essential in driving chemical processes. It can take various forms, such as heat, light, and electrical energy. During chemical reactions, bonds between atoms are broken and formed, which involves the absorption or release of energy.

Endothermic reactions absorb energy from the surroundings (e.g., photosynthesis in plants).

Exothermic reactions release energy (e.g., combustion of fuels).

Conservation of Mass and Energy: According to the Law of Conservation of Mass, matter cannot be created or destroyed in a chemical reaction. Similarly, the Law of Conservation of Energy states that energy can neither be created nor destroyed, only converted from one form to another.

6. Reactions and Equations

Chemical reactions are at the heart of chemistry. They describe the process by which substances interact to form new substances.

Reactants are the starting materials in a chemical reaction, and **products** are the substances formed as a result of the reaction.

Chemical Equations represent chemical reactions symbolically. For example, the reaction between hydrogen and oxygen to form water is written as:

2H2+O2→2H2O2H_2 + O_2 \rightarrow 2H_2O2H2+O2→2H2O

Balancing these equations is a critical skill in chemistry, as it ensures that the Law of Conservation of Mass is obeyed.

7. Chemistry as the Central Science

Chemistry is often called the "central science" because it connects and overlaps with several other scientific disciplines:

Physics: Chemistry and physics overlap in areas like thermodynamics, quantum mechanics, and atomic theory. The principles of energy, force, and motion in physics are essential for understanding chemical reactions and molecular behavior.

Biology: Biochemistry, a branch of chemistry, is crucial for understanding life processes. It explains how enzymes work, how DNA replicates, and how energy is produced and used by cells.

Environmental Science: Chemistry is essential for studying pollutants, water quality, and atmospheric processes. It helps in developing sustainable practices and technologies to combat environmental issues like climate change.

Medicine and Pharmacology: The development of drugs and medical treatments relies heavily on chemistry. Understanding how molecules interact with the human body allows for the design of new medications and therapies.

8. Applications of Chemistry in Daily Life

Chemistry's impact is far-reaching and integral to many aspects of daily life. Some practical applications include:

Food and Cooking: Chemical processes like fermentation, caramelization, and the Maillard reaction all play critical roles in food preparation and flavor development.

Cleaning and Hygiene: Household cleaning products rely on the principles of chemistry to break down dirt, grease, and stains. Soaps, detergents, and disinfectants are all products of chemical engineering.

Energy: Batteries, fuels, and renewable energy technologies are based on chemical principles. Understanding how energy is stored, transferred, and utilized is essential for developing sustainable energy solutions.

Materials Science: Chemistry is essential in the development of new materials, such as plastics, ceramics, and alloys. Innovations in chemistry have led to advances in technology, construction, and manufacturing.

Healthcare: From the synthesis of pharmaceuticals to the creation of medical devices, chemistry has revolutionized healthcare. Chemical reactions are behind the development of vaccines, diagnostic tools, and treatments for diseases.

In summary, chemistry is the science that explains the fundamental building blocks of the universe—matter, its properties, and its transformations. It is the study of the interactions between substances and the changes that arise from those interactions. As the "central science," chemistry connects with many other scientific fields and plays a pivotal role in understanding and shaping the world around us.

Whether it's in the air we breathe, the energy we use, or the products we consume, chemistry is at the core of almost every aspect of our lives. By studying chemistry, we gain insight into the workings of the natural world and acquire the knowledge needed to solve global challenges, from developing sustainable energy solutions to creating new medical treatments.

The Importance of Chemistry in Daily Life

Chemistry is often regarded as a distant and abstract science, but it is actually deeply embedded in every aspect of our daily lives. From the food we eat to the air we breathe, from the products we use to the natural processes that sustain life, chemistry plays an essential role in shaping and enhancing the quality of human life. The study of chemistry provides us with the knowledge and tools to understand and manipulate the substances and processes that are crucial to health, technology, industry, and the environment.

Let's explore how chemistry is involved in key areas of everyday life and why understanding it is so crucial.

1. Chemistry in Food and Cooking

Food, a fundamental necessity for life, is composed of a complex array of chemicals. The flavors, textures, and nutritional values of what we eat are all determined by chemical compounds and reactions. Whether we realize it or not, every time we cook, we are conducting a series of chemical transformations. These chemical processes are what give food its taste, appearance, and digestibility.

Nutritional Chemistry: Our bodies rely on the chemical breakdown of food to extract nutrients. Carbohydrates, proteins, and fats are digested through enzymatic reactions into simpler molecules like sugars, amino acids, and fatty acids. These are then absorbed and used for energy, growth, and repair.

Chemical Reactions in Cooking: Cooking is a practical application of chemistry. When we bake, roast, or fry food, chemical reactions like the Maillard reaction (which gives browned food its distinctive flavor) and caramelization take place. These reactions not only affect the flavor but also the texture and color of the food.

Food Preservation: Chemistry has revolutionized food preservation techniques. Salting, drying, and fermenting are ancient methods based on chemical principles. Modern techniques like freezing, canning, and the use of preservatives are all rooted in chemistry, helping prevent spoilage and extend the shelf life of food.

2. Chemistry in Health and Medicine

Chemistry is fundamental to our understanding of health, disease, and medicine. The pharmaceutical industry, which develops life-saving drugs and treatments, relies heavily on chemical research to synthesize and produce medications. Additionally, chemical processes underpin biological functions in the body, from metabolism to the immune response.

Pharmaceutical Chemistry: The development of medications and vaccines is one of the most direct applications of chemistry in healthcare. Chemists work to understand how drugs interact with biological systems, which enables the creation of medicines that can treat a wide range of diseases and conditions. For example, the development of antibiotics, antivirals, and vaccines is a product of advanced chemical research.

Biochemical Processes in the Body: Our bodies are essentially complex chemical systems. Every process that keeps us alive—from breathing to digestion to neural activity—involves chemical reactions. Hormones, enzymes, and neurotransmitters are all chemicals that regulate vital functions in the body.

Medical Diagnostics: Diagnostic tools like blood tests, MRIs, and X-rays rely on chemistry. Blood chemistry tests measure levels of different chemicals in the blood to assess organ function and detect diseases. The contrast dyes used in imaging technologies are based on chemical compounds designed to enhance the visibility of internal structures.

3. Chemistry in Cleaning and Hygiene

Chemistry plays a vital role in maintaining cleanliness and hygiene, which are crucial for preventing disease and promoting health. Cleaning products, disinfectants, and personal care items are formulated using chemical principles to target and remove contaminants, germs, and dirt.

Soaps and Detergents: Soaps and detergents are made from surfactants, which are chemical compounds that help break down oils, grease, and dirt, allowing them to be washed away with water. The chemical structure of these compounds has both hydrophilic (water-attracting) and hydrophobic (water-repelling) ends, making them effective at cleaning.

Disinfectants: Products like bleach, hand sanitizers, and antibacterial sprays are formulated using chemicals that kill or inhibit the growth of harmful microorganisms. These disinfectants work by disrupting the cell walls of bacteria or the protein structures of viruses, preventing them from reproducing.

Personal Care Products: Chemistry is also behind everyday products like shampoos, toothpaste, deodorants, and skincare items. These products are carefully designed to be safe and effective, combining various chemical ingredients to achieve desired effects, such as cleaning, moisturizing, or protecting against UV rays.

4. Chemistry in the Environment

Chemistry is essential for understanding environmental processes and addressing challenges such as pollution, climate change, and resource conservation. From the air we breathe to the water we drink, chemistry helps explain natural processes and human impact on the environment.

Air Quality and Pollution: The composition of the air, as well as the pollutants released into it, are all chemical in nature. Understanding atmospheric chemistry helps scientists study the impact of pollutants like carbon dioxide, sulfur dioxide, and nitrogen oxides, which contribute to air pollution and climate change. The development of technologies for cleaner energy, air purification, and emission control is rooted in chemistry.

Water Chemistry: Water purification, desalination, and treatment of wastewater are vital chemical processes that ensure safe drinking water and protect aquatic ecosystems. Chemistry is also central to understanding issues like acid rain, ocean acidification, and the effects of pollutants on marine life.

Green Chemistry: In response to environmental challenges, green chemistry has emerged as a field that focuses on designing chemical processes and products that reduce waste, minimize toxicity, and use renewable resources. By applying principles of green chemistry, industries can lower their environmental footprint and create sustainable solutions for the future.

5. Chemistry in Energy Production

Energy, in its various forms, is fundamentally linked to chemical processes. From traditional fossil fuels to renewable energy sources, chemistry is critical in developing and refining the ways we produce, store, and use energy.

Fossil Fuels and Combustion: The burning of fossil fuels (coal, oil, and natural gas) for energy involves chemical reactions where carbon-based compounds react with oxygen to produce carbon dioxide, water, and energy. Understanding these combustion reactions is essential for optimizing energy production and addressing the environmental impact of carbon emissions.

Battery Technology: Batteries rely on electrochemical reactions to store and release energy. Chemistry is crucial in developing new battery technologies, such as lithium-ion batteries, which power everything from smartphones to electric vehicles. Advances in battery chemistry are helping to create more efficient, longer-lasting energy storage solutions.

Renewable Energy: Chemistry is at the heart of renewable energy technologies. Solar panels, for example, use chemical reactions to convert sunlight into electricity. Hydrogen fuel cells, which generate electricity through chemical reactions between hydrogen and oxygen, are another example of chemistry-based energy innovation.

6. Chemistry in Materials and Technology

The materials we use in daily life—plastics, metals, textiles, and electronics—are all products of chemical innovation. Advances in materials science, which is a subfield of chemistry, have transformed industries ranging from construction to consumer electronics.

Plastics and Polymers: Chemistry has enabled the creation of synthetic materials like plastics and polymers, which have countless applications in packaging, construction, automotive manufacturing, and more. These materials are engineered at the molecular level to be durable, lightweight, and versatile.

Metals and Alloys: The development and refinement of metals and alloys (combinations of metals) rely on chemical processes. Chemistry allows for the creation of materials with specific properties, such as stainless steel for corrosion resistance or aluminum alloys for lightweight strength.

Semiconductors and Electronics: Modern electronics are built on semiconductors, materials that have electrical conductivity between that of a conductor and an insulator. Semiconductor technology, which is based on principles of chemistry, is critical for the development of transistors, microchips, and other components that power computers, smartphones, and other electronic devices.

7. Chemistry in Art and Culture

Even in the realms of art and culture, chemistry plays an important role. From the pigments used in paintings to the materials used in sculpture, and even the preservation of historical artifacts, chemistry is key.

Paints and Pigments: The colors in paints and dyes are derived from various chemical compounds. Chemists create pigments by manipulating the molecular structures of substances to produce vibrant and lasting colors. Advances in chemistry have expanded the range of available colors and improved the durability of paints.

Preservation of Artifacts: Museums and conservationists use chemistry to preserve historical artifacts and works of art. By understanding the chemical composition of materials, they can develop methods to prevent degradation, such as controlling humidity levels to preserve paintings or using chemical treatments to stabilize ancient manuscripts.

In conclusion, chemistry is deeply woven into the fabric of daily life. It affects nearly every aspect of human existence, from the food we eat to the technologies we use, from our health and well-being to the environment we live in. Understanding the principles of chemistry not only helps us appreciate the world around us but also empowers us to solve some of the most pressing challenges facing humanity, such as energy production, environmental protection, and medical advancements.

By studying chemistry, we gain insights into how the universe operates at a molecular level, allowing us to harness its power to improve our lives and create a better future. Whether we realize it or not, chemistry is at the heart of innovation, health, and the environment, making it one of the most essential sciences for modern life.

The History of Chemistry

The history of chemistry is a fascinating journey that spans thousands of years, from the early civilizations' rudimentary understanding of materials and substances to the complex and sophisticated science we know today. Chemistry has its roots in ancient alchemy but evolved into a distinct scientific discipline through a combination of philosophical thought, experimentation, and technological advances.

This development reflects humanity's desire to understand and manipulate the matter around us, from the air we breathe to the metals we shape, and from the medicines we use to the materials that build our modern world. Let’s explore the major periods and milestones in the history of chemistry, highlighting key figures, discoveries, and transformations that shaped the field.

1. Ancient Chemistry: Alchemy and Early Theories

The earliest precursors of modern chemistry can be found in the practices of alchemy, metallurgy, and natural philosophy. Ancient civilizations, such as those in Egypt, Mesopotamia, China, India, and Greece, had a rudimentary understanding of chemical processes, mainly through practical applications like metallurgy, pottery, and medicine. However, these early efforts were not based on scientific principles but rather on trial and error, observation, and mysticism.

Metallurgy and Practical Chemistry: In ancient times, people learned to extract metals from ores and develop techniques to refine and shape them. For example, the Egyptians mastered the art of extracting and working with gold, copper, and iron. The production of alloys, such as bronze (a mixture of copper and tin), marked a significant advancement, leading to the Bronze Age around 3300 BCE.

Alchemy: Alchemy was practiced across different cultures, particularly in Egypt, Greece, the Islamic world, China, and India. Alchemists sought to transform base metals into noble metals like gold and silver, a pursuit that became known as the "philosopher’s stone." While alchemy was largely mystical and often steeped in religious or spiritual overtones, it laid the groundwork for experimental practices that would later be refined into chemistry.

In Egypt, alchemists like **Zosimos of Panopolis** (circa 300 CE) wrote extensively about the processes of distillation, calcination, and sublimation—methods still used in modern chemistry.

In China, alchemists pursued immortality elixirs and experimented with substances such as mercury and sulfur. These efforts contributed to early discoveries in the properties of substances, though they were not understood in a scientific context.

Greek Natural Philosophy: The ancient Greeks made significant theoretical contributions to early chemistry. Philosophers like **Empedocles** (495–435 BCE) proposed that all matter was composed of four fundamental elements: earth, air, fire, and water. Later, **Democritus** (circa 460–370 BCE) introduced the concept of the atom, positing that all matter was composed of indivisible particles called "atoms" (from the Greek word "atomos," meaning uncuttable). Although Democritus' atomic theory was speculative and lacked experimental proof, it foreshadowed modern atomic theory.

2. The Islamic Golden Age and the Birth of Chemistry

During the Islamic Golden Age (8th to 14th centuries), scholars in the Islamic world preserved, translated, and expanded upon the works of the Greeks and other ancient civilizations. These scholars made significant advancements in alchemy, laying the groundwork for modern chemistry through experimentation, meticulous documentation, and the introduction of new scientific methods.

Jabir ibn Hayyan (Geber): Often considered the "father of chemistry," Jabir ibn Hayyan (circa 721–815 CE) was an alchemist and polymath who made substantial contributions to early chemical knowledge. His works included descriptions of distillation, crystallization, and the preparation of various chemicals like acids. Jabir's systematic approach to chemical processes emphasized the importance of observation and experimentation, distinguishing his work from the more mystical aspects of alchemy.

Al-Razi (Rhazes): Another key figure from the Islamic world was **Al-Razi** (865–925 CE), a Persian polymath who made advancements in medical chemistry. He wrote extensively about substances like alcohol and sulfuric acid and conducted experiments that were among the first to approach chemistry from a more empirical and scientific perspective.

Ibn Sina (Avicenna): Ibn Sina (980–1037 CE), a Persian physician and philosopher, also contributed to the development of chemistry, particularly in the realm of medicine and pharmacology. He conducted detailed studies of distillation and purification processes that would later influence European chemists during the Renaissance.

The contributions of Islamic scholars helped shape the knowledge base of chemistry, and their works were later translated into Latin and spread throughout Europe, leading to the Renaissance.

3. The Renaissance and the Transition to Modern Chemistry

The European Renaissance (14th to 17th centuries) saw the revival of classical knowledge and a growing interest in the natural sciences, including chemistry. This period marked the transition from alchemy to a more systematic and scientific approach to studying matter. New discoveries in physics, biology, and astronomy also influenced the development of chemistry during this era.

Paracelsus (1493–1541): A Swiss physician and alchemist, **Paracelsus** challenged the mystical aspects of alchemy and emphasized the use of chemicals and minerals in medicine. He is credited with founding the field of medical chemistry and introducing the use of chemical compounds in treating diseases. Paracelsus' focus on experimentation and the practical application of chemicals in medicine helped shift chemistry towards a more scientific and empirical discipline.

Robert Boyle (1627–1691): Often considered the "father of modern chemistry," **Robert Boyle** was an English natural philosopher who made significant contributions to the scientific method and the study of gases. His most famous work, *The Sceptical Chymist* (1661), challenged the classical elements of earth, air, fire, and water and argued that matter was composed of atoms and chemical elements. Boyle's experiments led to the formulation of **Boyle's Law**, which describes the inverse relationship between the pressure and volume of a gas. He also advocated for rigorous experimental methods, helping to distinguish chemistry as a science separate from alchemy.

4. The Age of Enlightenment and the Birth of Modern Chemistry

The 18th century, known as the Age of Enlightenment, was a time of great intellectual and scientific advancement. During this period, chemistry began to take its modern form, characterized by a focus on quantitative experiments, the identification of chemical elements, and the rejection of alchemical mysticism.

Antoine Lavoisier (1743–1794): Widely regarded as the "father of modern chemistry," **Antoine Lavoisier** revolutionized the field by introducing the concept of the conservation of mass in chemical reactions. He demonstrated that during a chemical reaction, matter is neither created nor destroyed, only transformed—a principle now known as the **Law of Conservation of Mass**. Lavoisier also named and characterized many chemical elements, such as oxygen, hydrogen, and sulfur, and disproved the phlogiston theory (an early attempt to explain combustion). His work laid the foundation for modern chemical nomenclature and the scientific method in chemistry.

Joseph Priestley (1733–1804) and **Carl Wilhelm Scheele (1742–1786)**: These two scientists, working independently, made important discoveries in gases. **Priestley** is credited with discovering oxygen in 1774, although **Scheele** had isolated it earlier but did not publish his findings until later. Priestley's and Scheele's work on gases contributed to the understanding of chemical reactions involving air and combustion, helping to dismantle the phlogiston theory.

John Dalton (1766–1844): Building on the work of earlier scientists, **John Dalton** developed the first modern atomic theory in the early 19th century. He proposed that matter is made up of indivisible atoms, and that each element is composed of atoms of a specific kind. Dalton's atomic theory provided a framework for understanding chemical reactions as the rearrangement of atoms, and it became one of the cornerstones of modern chemistry.

5. The 19th Century: The Rise of Chemical Science

The 19th century saw chemistry emerge as a well-defined scientific discipline. Major advancements were made in understanding the composition of matter, the periodic organization of elements, and the structure of molecules.

Dmitri Mendeleev (1834–1907): One of the most significant achievements of 19th-century chemistry was the development of the **Periodic Table of Elements** by **Dmitri Mendeleev** in 1869. Mendeleev arranged the known elements based on their atomic weights and chemical properties, revealing periodic trends in the behavior of elements. His periodic table not only organized the elements but also predicted the existence and properties of elements that had not yet been discovered.

Amedeo Avogadro (1776–1856): **Avogadro's Law**, formulated in 1811, stated that equal volumes of gases, at the same temperature and pressure, contain an equal number of molecules. This principle was critical in advancing the understanding of molecular structure and atomic theory.

Organic Chemistry: The 19th century also saw the rise of **organic chemistry**, the study of carbon-containing compounds. **Friedrich Wöhler**'s synthesis of urea in 1828 from inorganic compounds marked the first time an organic compound had been created from non-organic materials, challenging the notion that organic compounds could only be produced by living organisms. This breakthrough led to the development of the field of organic chemistry, which would later include the study of hydrocarbons, polymers, and pharmaceuticals.

6. The 20th Century and Beyond: Modern Chemistry

The 20th century saw chemistry evolve into a highly specialized and advanced science, with major discoveries in atomic structure, quantum mechanics, and biochemistry.

Marie Curie (1867–1934) and **Radioactivity**: **Marie Curie** and her husband **Pierre Curie** conducted groundbreaking research on radioactivity, leading to the discovery of elements like polonium and radium. Marie Curie's work laid the foundation for nuclear chemistry and the study of radioactive elements, which would later have significant implications for energy production and medical treatments.

Quantum Chemistry: In the early 20th century, the development of quantum mechanics transformed the understanding of atomic and molecular structures. Scientists like **Niels Bohr**, **Erwin Schrödinger**, and **Werner Heisenberg** developed models that explained the behavior of electrons in atoms, which in turn allowed for more accurate predictions of chemical reactions and bonding.

Molecular Biology and Biochemistry: The discovery of the structure of DNA by **James Watson** and **Francis Crick** in 1953, along with the elucidation of the genetic code, bridged chemistry with biology. This marked the beginning of modern **biochemistry** and **molecular biology**, fields that study the chemical processes of life and have led to advances in genetics, medicine, and biotechnology.

Green Chemistry: In recent decades, the field of chemistry has increasingly focused on sustainability and environmental impact. **Green chemistry** aims to design chemical processes and products that reduce or eliminate hazardous substances, minimizing the environmental footprint of chemical industries.

The history of chemistry is a rich and complex narrative that spans from ancient alchemy to modern scientific advances. Each era, from early metallurgists to modern quantum chemists, contributed to our current understanding of matter, its properties, and the reactions that transform it. As chemistry continues to evolve, it remains one of the most dynamic and essential sciences, driving innovation and addressing global challenges such as energy, health, and the environment.

Branches of Chemistry

Chemistry is an expansive and diverse field of science that explores the composition, structure, properties, and changes of matter. Over time, this vast domain has been divided into specialized branches, each focusing on particular aspects of chemistry or areas of application. These branches of chemistry enable scientists to address specific problems, understand complex systems, and create new technologies. Here, we will explore the primary branches of chemistry, outlining their scope, importance, and key areas of research.

1. Organic Chemistry

Organic chemistry is the branch of chemistry that focuses on the study of carbon-containing compounds. Given that carbon atoms can form stable bonds with a variety of other elements, especially hydrogen, oxygen, nitrogen, and sulfur, organic chemistry is central to the understanding of life processes and a wide range of synthetic materials.

Scope: Organic chemistry primarily deals with hydrocarbons (compounds made up of hydrogen and carbon) and their derivatives. It also studies the structure, properties, composition, reactions, and preparation of organic compounds. Organic chemistry is essential in the study of living organisms, as it is the foundation for biochemistry and molecular biology.

Applications: Organic chemistry plays a crucial role in numerous industries, including pharmaceuticals, petrochemicals, agriculture, and plastics. It is responsible for the development of drugs, dyes, polymers, and many other materials. Organic chemists study reactions like polymerization, oxidation, and substitution, creating compounds like aspirin, polyethylene, and synthetic rubber.

Key Topics: Functional groups (like alcohols, amines, and carboxylic acids), isomerism, stereochemistry, and organic reactions (such as addition, elimination, and nucleophilic substitution).

2. Inorganic Chemistry

Inorganic chemistry is the study of inorganic compounds, which are generally compounds that do not contain carbon-hydrogen bonds (with some exceptions like carbonates, cyanides, and oxides). This branch covers all chemical compounds outside of organic compounds, focusing primarily on metals, minerals, and non-metal elements.

Scope: Inorganic chemistry involves the study of metals, salts, and coordination compounds, as well as materials such as ceramics, metals, and catalysts. It also investigates the behavior of inorganic elements in different chemical environments, including how they react with other elements and compounds.

Applications: Inorganic chemistry is vital in materials science, catalysis, metallurgy, and environmental science. It is key to developing materials like semiconductors, pigments, fertilizers, and superconductors. Inorganic chemists work on compounds that are critical in industries such as mining, electronics, and agriculture.

Key Topics: Coordination chemistry, crystal field theory, transition metals, non-metals, bioinorganic chemistry, and inorganic reactions such as redox reactions and acid-base chemistry.

3. Physical Chemistry

Physical chemistry is the branch that deals with the application of physics to chemical systems. It focuses on understanding the physical properties of molecules, the forces that act upon them, and how these properties relate to chemical reactions.

Scope: Physical chemistry uses principles of thermodynamics, quantum mechanics, kinetics, and statistical mechanics to explain and predict the behavior of chemical systems. It addresses how energy changes during reactions, the rates of these reactions, and how matter behaves on a molecular and atomic scale.

Applications: Physical chemistry is essential in fields such as material science, nanotechnology, and pharmaceuticals. It helps in understanding reaction mechanisms, energy transfer, and molecular dynamics, which are critical in designing new materials, understanding catalysis, and improving energy storage systems. It is also foundational for developing new theories and models in chemistry.

Key Topics: Thermodynamics, quantum chemistry, chemical kinetics, spectroscopy, and surface chemistry.

4. Analytical Chemistry

Analytical chemistry involves the analysis of substances to determine their composition and quantity. It focuses on the separation, identification, and quantification of the chemical components of natural and artificial materials.

Scope: Analytical chemistry provides tools and methods to analyze chemical compounds and determine their concentration in a given sample. It is central to quality control, environmental monitoring, food safety, and forensic science. Analytical chemists use a wide array of techniques, including spectroscopy, chromatography, and electrochemical analysis.

Applications: Analytical chemistry is used in pharmaceutical testing, environmental analysis (e.g., detecting pollutants in air, water, and soil), and quality assurance in industries like food production and manufacturing. Analytical chemists develop methods for detecting trace amounts of substances, identifying unknown compounds, and monitoring chemical processes in real time.

Key Topics: Quantitative and qualitative analysis, chromatography, mass spectrometry, titration, electrochemical analysis, and spectroscopy.

5. Biochemistry

Biochemistry is the branch of chemistry that studies the chemical processes within and related to living organisms. It is an interdisciplinary field that bridges biology and chemistry, focusing on the molecular level of biological processes.

Scope: Biochemistry investigates the structures, functions, and interactions of biological molecules such as proteins, lipids, carbohydrates, and nucleic acids. It plays a crucial role in understanding cellular processes, metabolism, genetics, and diseases. Biochemistry examines how these molecules interact in processes like replication, transcription, translation, and cellular respiration.

Applications: Biochemistry is fundamental to medicine, biotechnology, and pharmacology. It helps in understanding diseases, developing new drugs, and designing therapies for genetic disorders. Biochemists study enzymes, hormones, and metabolic pathways to understand their roles in health and disease. The development of vaccines, diagnostic tools, and genetically modified organisms (GMOs) are direct outcomes of biochemistry research.

Key Topics: Enzymes, metabolism, molecular genetics, bioenergetics, and the structure-function relationship of biomolecules.

6. Environmental Chemistry

Environmental chemistry focuses on the chemical phenomena occurring in the environment. It deals with the chemical processes and interactions that govern natural and human-induced environmental changes.

Scope: This branch of chemistry studies the chemical composition of the air, water, and soil, and how pollutants and other chemical agents affect ecosystems. Environmental chemistry also examines the movement of chemical substances through the environment, including the cycling of elements such as carbon, nitrogen, and phosphorus. The study of pollutants, their impact on the environment, and the development of methods for pollution control and remediation are central to environmental chemistry.

Applications: Environmental chemists work on issues like air and water pollution, hazardous waste management, and climate change. They develop cleaner industrial processes (green chemistry), monitor and manage pollution levels, and develop strategies for sustainable development. Environmental chemistry is also crucial for understanding and mitigating the effects of chemical spills, pesticide use, and other human activities on ecosystems.

Key Topics: Pollution, the carbon cycle, chemical degradation, green chemistry, atmospheric chemistry, and environmental monitoring.

7. Industrial Chemistry

Industrial chemistry applies chemical principles to large-scale manufacturing processes. It focuses on converting raw materials into valuable products through chemical reactions and processes.

Scope: Industrial chemistry involves the design, optimization, and management of chemical processes used in industries such as petroleum refining, pharmaceuticals, polymers, fertilizers, and food production. It bridges chemistry with engineering and economics to ensure that chemical processes are efficient, sustainable, and economically viable.

Applications: Industrial chemists are involved in creating fuels, materials, and chemicals used in everyday life. They also work on improving the efficiency of industrial processes, reducing environmental impact, and developing new technologies such as biodegradable plastics and renewable energy sources. Industrial chemistry plays a critical role in the global economy, from producing gasoline and fertilizers to manufacturing everyday products like soaps, plastics, and textiles.

Key Topics: Chemical engineering, process design, catalysis, materials chemistry, and production techniques.

8. Theoretical Chemistry

Theoretical chemistry uses mathematical models and simulations to understand and predict chemical phenomena. It is a highly specialized field that combines principles from chemistry and physics to explain the behavior of atoms and molecules.

Scope: Theoretical chemists use computational techniques and quantum mechanics to model molecular behavior, reaction mechanisms, and chemical properties. This branch focuses on predicting the outcomes of chemical reactions, understanding molecular structures, and designing new materials and drugs through simulation rather than experimentation.

Applications: Theoretical chemistry is crucial in drug design, materials science, and nanotechnology. It allows scientists to simulate the properties of new compounds before they are synthesized in the laboratory, saving time and resources. Theoretical chemists also contribute to developing new chemical theories and refining existing models to better understand chemical systems.

Key Topics: Quantum chemistry, molecular dynamics, computational chemistry, and statistical mechanics.

Each branch of chemistry contributes to a different aspect of our understanding of matter and its interactions. Whether it's creating new medicines, developing sustainable materials, understanding biological processes, or analyzing environmental pollutants, the branches of chemistry are essential for solving some of the world's most pressing problems. These branches, while distinct in their focus, are often interconnected, allowing for cross-disciplinary approaches and innovations that push the boundaries of science. By studying and mastering the various branches of chemistry, scientists can continue to make significant advancements that improve our daily lives and address global challenges.

The Scientific Method and its Application in Chemistry

The **scientific method** is a systematic, logical approach used by scientists to explore observations, answer questions, and test hypotheses. In chemistry, this method is particularly crucial as it provides a structured process for experimentation and discovery, allowing chemists to develop a deeper understanding of the properties, behaviors, and interactions of matter.

The scientific method is not a rigid set of steps but rather a flexible framework that scientists use to guide their research. By using this method, chemists can minimize biases, errors, and subjective influences, ensuring that their findings are objective, reproducible, and reliable. The following sections will detail the components of the scientific method and explore how it is applied in chemistry.

Key Steps in the Scientific Method

- Observation
- Question
- Hypothesis Formation
- Experimentation
- Data Analysis
- Conclusion
- Communication of Results

1. Observation

Observation is the first step in the scientific method and involves noticing phenomena or patterns that spark curiosity or raise questions. In chemistry, observations can be either qualitative (descriptive, non-numerical) or quantitative (measurable, numerical). For example:

Qualitative observation: A chemist might observe that a metal turns dull over time when exposed to air (rust formation).

Quantitative observation: A chemist might measure the rate at which a reaction proceeds, such as how fast a certain amount of reactant is consumed.

Observations in chemistry often lead to identifying problems or gaps in knowledge. For example, a chemist might observe that certain chemicals behave differently under specific conditions, prompting further inquiry into why this happens.

2. Question

After making observations, the next step is to ask a question. In chemistry, questions usually arise from observed phenomena that require explanation or investigation. A good scientific question is specific, measurable, and testable through experimentation.

For example, after observing that iron rusts when exposed to moisture, a chemist might ask: **"What chemical process causes iron to rust?"** or **"What factors accelerate the rusting of iron?"**

3. Hypothesis Formation

A **hypothesis** is an educated guess or a tentative explanation that answers the question posed. It must be testable and falsifiable, meaning that experiments can either support or disprove it. In chemistry, a hypothesis often predicts how a chemical reaction will occur or how changing a variable will affect the outcome of an experiment.

For instance, a chemist might hypothesize: **"Iron rusts because it reacts with oxygen and water to form iron oxide."** This hypothesis suggests a potential chemical process and can be tested by conducting experiments under controlled conditions.

In chemistry, hypotheses are sometimes developed using prior knowledge from chemical principles, such as atomic theory, molecular structure, or bonding behavior.

4. Experimentation

Experiments are the heart of the scientific method. They are carefully designed procedures to test the hypothesis. In chemistry, experimentation involves manipulating one or more variables (independent variables) while keeping others constant (controlled variables) to observe the effect on a dependent variable (e.g., the rate of a chemical reaction or the formation of a product).

Chemists often conduct **controlled experiments**, where they compare results from an experimental group to a control group. For example, to test whether moisture accelerates rusting, a chemist might set up one experiment with dry iron and another with wet iron under identical conditions.

Designing an experiment: This involves selecting appropriate reagents, setting up laboratory equipment, and defining measurable outcomes (e.g., mass changes, color changes, gas production).

Reproducibility: One of the key aspects of scientific experiments is reproducibility. Other scientists should be able to replicate the experiment and obtain similar results under the same conditions. This ensures that the findings are not due to chance or error.

In chemistry, experiments can vary from simple reactions performed in test tubes to highly sophisticated processes using advanced technologies like spectroscopy, chromatography, or X-ray crystallography.

5. Data Collection and Analysis

Once the experiment is performed, chemists gather data, which can be qualitative or quantitative. In chemistry, quantitative data is often emphasized because it provides precise and measurable information about chemical processes.

For example, a chemist may measure:

The temperature at which a reaction occurs.

The amount of product formed during a reaction.

The concentration of reactants before and after the reaction.

Data analysis often involves calculating results, plotting graphs, and using statistical methods to determine trends or relationships. A chemist may analyze how changes in temperature affect the reaction rate or how the concentration of a solution influences the yield of a chemical product.

For more complex chemical processes, chemists may use computational models or software to analyze data and predict future outcomes.

6. Conclusion

Based on the data, the chemist draws a conclusion. The conclusion must directly address the hypothesis, indicating whether the data supports or contradicts the original prediction. If the hypothesis is supported, it may lead to further research to expand on the findings. If the hypothesis is not supported, it may need to be revised, and new experiments designed.

For example, after observing that iron in a moist environment rusts faster than dry iron, a chemist may conclude that **moisture accelerates the oxidation process**, thus supporting the hypothesis that water and oxygen are both required for rust formation.

In some cases, experiments may lead to unexpected results, which can offer new insights or open up entirely new avenues of research.

7. Communication of Results

The final step in the scientific method is to share the results with the scientific community. This is often done through the publication of research papers in scientific journals, presentations at conferences, or even discussions within laboratory groups.

In chemistry, communicating results is critical because it allows others to review, critique, and replicate the findings. Peer review ensures the validity and accuracy of the results before they are widely accepted by the scientific community.

Chemists typically write detailed reports that include:

The background and rationale for the study.

The methods and materials used.

The experimental results and data.

Analysis and interpretation of the data.

A conclusion that ties back to the original hypothesis.

Sharing results also fosters collaboration and innovation, allowing other researchers to build upon previous findings.

Application of the Scientific Method in Chemistry

1. Developing New Materials

The scientific method is used extensively in the development of new materials. For example, chemists working in polymer chemistry may observe that certain materials degrade too quickly under heat. They can hypothesize that adding specific stabilizing agents will enhance the material's heat resistance. Through a series of controlled experiments, they test different additives, analyze the results, and refine their materials until they achieve the desired properties.

2. Pharmaceutical Research

In drug discovery, the scientific method plays a crucial role in identifying new therapeutic compounds. For example, researchers may observe that certain natural compounds have anti-inflammatory properties. Based on this, they might hypothesize that similar synthetic compounds could provide the same effect. By conducting experiments on these synthetic compounds, measuring their efficacy, and analyzing the results, chemists can determine whether a new drug candidate is worth pursuing.

3. Environmental Chemistry

The scientific method is also applied in environmental chemistry to understand pollution and develop remediation strategies. A chemist might observe that a particular area of soil is contaminated with heavy metals. They could hypothesize that adding a specific chemical treatment will neutralize the pollutants. Experiments are conducted to test the treatment's effectiveness, and data is collected to assess its impact on the environment.

The scientific method is fundamental to the discipline of chemistry, guiding researchers through the process of inquiry, experimentation, and discovery. By applying this method, chemists can develop theories, uncover new knowledge, and find practical solutions to real-world problems. Whether creating new materials, discovering drugs, or addressing environmental issues, the scientific method ensures that chemistry remains a precise, logical, and reliable science.

Safety in the Chemistry Laboratory

Safety is paramount in any chemistry laboratory, where hazardous substances, potentially dangerous reactions, and sensitive equipment are regularly encountered. Proper understanding and application of laboratory safety protocols protect both the individuals conducting experiments and those around them, while ensuring that research and educational activities proceed smoothly without accidents.

The importance of safety in the chemistry lab cannot be overstated, as even seemingly minor mistakes or lapses in judgment can lead to serious injuries, environmental damage, or compromised research. This section provides an extensive discussion on key safety principles, protocols, and equipment necessary for a safe and efficient chemistry laboratory experience.

Importance of Laboratory Safety

Preventing Injuries and Accidents: Chemicals can be corrosive, toxic, flammable, or reactive. Proper handling ensures that accidents like spills, fires, and explosions are avoided, minimizing the risk of harm to people.

Protecting the Environment: Laboratory safety procedures also extend to how chemical waste is disposed of. Safe disposal prevents pollution, contamination, and long-term environmental harm.

Preserving Research Integrity: Adherence to safety protocols ensures that experiments proceed without incidents that could alter results, damage equipment, or halt important research.

Key Elements of Laboratory Safety

1. Personal Protective Equipment (PPE)

Personal protective equipment is the first line of defense in ensuring laboratory safety. Different types of PPE are designed to protect specific areas of the body from chemical exposure, burns, cuts, or other injuries.

Safety Goggles: Protects the eyes from chemical splashes, flying debris, or hazardous fumes. Ordinary glasses do not provide sufficient protection.

Lab Coats: Provides a barrier between chemicals and the skin or personal clothing. Lab coats should be flame-resistant and made from non-reactive materials.

Gloves: Different gloves are made from various materials (latex, nitrile, neoprene, etc.) and should be selected based on the type of chemicals being handled. They protect hands from corrosive chemicals, toxic substances, and cuts.

Closed-Toe Shoes: To protect feet from chemical spills or falling objects, shoes must cover the entire foot. Sandals or open-toed shoes are never allowed.

Respirators or Masks: In cases where fumes, dust, or gases are present, respirators or chemical masks may be necessary to prevent inhalation of harmful substances.

Face Shields: These offer additional protection for the face and neck during procedures that might result in splashes or explosions.

2. Proper Handling and Storage of Chemicals

Chemicals in a laboratory can range from harmless to highly dangerous. Understanding the properties of each chemical and how to handle and store them properly is critical for safety.

Labeling: All chemicals in the lab must be properly labeled with their name, concentration, and any hazard warnings (e.g., flammable, corrosive, toxic). Mislabeling can lead to dangerous mix-ups.

Storage: Chemicals should be stored according to their hazard class. For instance:

Flammable liquids like ethanol or acetone should be stored in flame-proof cabinets.

Acids and bases should be kept separately to prevent dangerous reactions if spilled or mixed.

Reactive chemicals (e.g., sodium or potassium) that react with water or air must be stored in a dry, inert environment.

Toxic or carcinogenic chemicals should be stored in well-ventilated areas, often with additional security measures to limit exposure.

Handling: Chemicals should always be handled with care. This includes pouring or mixing under a fume hood if fumes are hazardous, using appropriate containers, and wearing gloves. Never pour water into concentrated acids (always add acid to water slowly) as it can cause splattering and releases heat.

3. Chemical Waste Disposal

Improper disposal of chemical waste can result in environmental harm, laboratory contamination, and dangerous chemical reactions. Laboratories must adhere to strict waste management protocols.

Segregation of Waste: Chemical waste should be separated into categories (e.g., organic, inorganic, acidic, basic, or toxic waste). Incompatible chemicals should never be disposed of together as they may react violently.

Disposal Containers: Special waste containers are designated for different types of waste, and these containers must be properly labeled and sealed when not in use. For example, flammable solvents must be stored in flame-proof waste bins.

Neutralization and Treatment: Some chemical waste (e.g., strong acids or bases) must be neutralized before disposal. This process must be done according to strict guidelines to prevent dangerous reactions.

4. Safe Laboratory Practices

Beyond personal protective equipment and chemical handling, good laboratory habits contribute to safety. These practices ensure that the laboratory environment remains orderly, efficient, and accident-free.

No Eating or Drinking: Food and beverages are prohibited in the lab to prevent accidental ingestion of chemicals. Additionally, food and drink could become contaminated by airborne chemicals or substances on lab surfaces.

Ventilation and Fume Hoods: Fume hoods are used to contain hazardous fumes, dust, and vapors generated during chemical reactions. Always perform reactions that involve volatile or harmful chemicals inside a fume hood to prevent inhalation.

Chemical Hygiene Plan: Laboratories often have a **Chemical Hygiene Plan** (CHP) that outlines proper procedures for handling chemicals, exposure prevention, and emergency procedures. All lab personnel must be familiar with this plan.

Avoiding Direct Contact: Always use equipment such as tongs, spatulas, or pipettes to handle chemicals, especially corrosive or toxic substances. Direct contact, even with gloves, should be minimized.

Clean Workspaces: Maintaining an organized and clean work environment reduces the risk of accidents like spills, cross-contamination, or trips and falls. Always clean up after experiments and dispose of waste appropriately.

Transporting Chemicals: When transporting chemicals within the lab, ensure that they are in sealed, shatter-proof containers. Larger containers should be transported using carts with spill containment measures.

5. Fire Safety

Many chemicals in a laboratory are flammable, making fire safety a critical concern. Fire prevention strategies and preparedness can prevent small incidents from escalating into dangerous situations.

Fire Extinguishers: Laboratories are equipped with different types of fire extinguishers based on the type of fire (e.g., Class B for flammable liquids or Class C for electrical fires). All personnel should know the location of fire extinguishers and how to use them.

Fire Blankets: These are used to smother small fires on people or equipment. Fire blankets can also help in case of clothing catching fire, as they can be wrapped around the individual to put out flames.

Emergency Exits: Labs must have clearly marked emergency exits and evacuation plans in case of a fire. Pathways to these exits should always be clear of obstructions.

No Open Flames: Avoid using open flames (e.g., Bunsen burners) unless absolutely necessary. Many chemicals are highly flammable and can ignite with the smallest spark.

6. Emergency Preparedness

Even with the most diligent safety protocols, accidents can happen. Therefore, being prepared for emergencies is essential to minimize injury and damage.

Eyewash Stations and Safety Showers: These are critical for quickly flushing chemicals from the body. Eyewash stations are used to rinse the eyes in case of splashes, while safety showers are used to wash off large spills or chemical exposure on the body. All personnel must know their locations and how to use them.

First Aid Kits: Laboratories must have easily accessible first aid kits stocked with materials to handle minor injuries (like cuts or burns) and chemical exposure (e.g., neutralizing agents).

Spill Kits: Spill kits are essential for cleaning up chemical spills safely. They typically include absorbents, neutralizing agents, gloves, and waste disposal bags. Different kits are available for acids, bases, solvents, and other hazardous materials.

Emergency Contact Information: Every lab should have posted emergency contact information for local emergency services, fire departments, and poison control centers.

7. Training and Documentation

Proper training is essential for all individuals working in a chemistry laboratory. Training ensures that lab personnel are familiar with safety protocols, chemical handling procedures, and emergency measures.

Safety Training: All lab personnel must undergo regular safety training sessions, covering topics like PPE usage, chemical handling, waste disposal, fire safety, and emergency response.

Standard Operating Procedures (SOPs): Laboratories often have written Standard Operating Procedures for using specific chemicals, conducting certain experiments, and handling equipment. These SOPs should be followed precisely to avoid accidents.

Material Safety Data Sheets (MSDS): MSDS (or Safety Data Sheets, SDS) provide detailed information about each chemical, including its hazards, handling precautions, storage requirements, and emergency measures in case of exposure. Chemists should always consult the SDS before using a new chemical.

Safety in the chemistry laboratory is a multifaceted discipline that requires constant vigilance, proper equipment, and thorough training. By adhering to safety protocols, utilizing personal protective equipment, properly handling chemicals, and preparing for emergencies, chemists can ensure that their work is conducted in a safe and controlled environment.

Ultimately, a safe laboratory fosters innovation, enhances productivity, and protects individuals and the environment from the potential hazards of chemical research and experimentation.

Basic Laboratory Equipment and Techniques

In the field of chemistry, the use of laboratory equipment and techniques is fundamental to conducting experiments, analyzing substances, and discovering new chemical phenomena. Whether in an academic setting or an industrial research laboratory, familiarity with basic laboratory tools and methods is essential for the accurate execution of experiments and the safe handling of chemicals.

Understanding the function and proper use of these instruments ensures efficiency, precision, and safety. This section provides an extensive overview of some of the most common pieces of laboratory equipment, their uses, and essential techniques that form the backbone of experimental chemistry.

1. Basic Laboratory Equipment

Laboratory equipment can range from simple glassware to sophisticated analytical instruments. Here are some of the most essential tools that chemists use on a daily basis.

a. Beakers

Description: Beakers are cylindrical glass containers with a flat bottom and a spout for pouring. They come in various sizes (e.g., 50 mL to 1000 mL) and are used for mixing, stirring, heating, and holding liquids.

Use: Beakers are essential for general-purpose mixing of chemicals and solutions. Though not highly precise for measuring volumes, they are used when approximate quantities are acceptable.

b. Erlenmeyer Flasks

Description: These are conical flasks with a narrow neck and flat base. The shape allows for easy mixing by swirling without spilling.

Use: Erlenmeyer flasks are commonly used for titrations, heating solutions, and storing liquids. The narrow neck can be stoppered to reduce evaporation or contamination.

c. Graduated Cylinders

Description: Graduated cylinders are tall, cylindrical containers marked with graduation lines for measuring liquids accurately.

Use: Unlike beakers or flasks, graduated cylinders are used when precise measurement of liquid volumes is required.

d. Burettes

Description: Burettes are long, graduated tubes with a stopcock at the bottom to release liquids in a controlled manner.

Use: They are essential in titrations, where a precise volume of one solution is added to another until a chemical reaction is completed. This is particularly important for quantitative analysis.

e. Pipettes

Description: Pipettes are slender tubes used to transport a measured volume of liquid from one container to another. They come in different types, including volumetric, graduated, and micropipettes.

Use: Pipettes are used for precision transfer of small quantities of liquids. Micropipettes are often employed in experiments requiring micro-liter volumes, such as in analytical chemistry and biology.

f. Volumetric Flasks

Description: These flasks have a round bottom and a long neck with a single graduation mark for measuring a precise volume.

Use: Volumetric flasks are used to prepare solutions of precise concentrations by dissolving a solute and diluting it to a known volume.

g. Test Tubes and Test Tube Racks

Description: Test tubes are small cylindrical tubes made of glass or plastic, usually used for holding small samples or conducting reactions on a small scale.

Use: Test tubes are used for performing simple qualitative experiments, such as observing reactions, heating small volumes of liquids, or holding biological samples.

h. Crucibles

Description: Crucibles are small, cup-shaped containers made of materials that can withstand high temperatures, such as porcelain or metal.

Use: They are used for heating substances to very high temperatures, especially for chemical reactions that require the material to be melted or strongly heated.

i. Mortar and Pestle

Description: A mortar is a bowl, and a pestle is a heavy club-shaped tool used to grind substances.

Use: In chemistry, mortar and pestles are used for grinding solid substances into fine powders, an important step in preparing reagents or catalysts.

j. Balances

Description: Analytical balances are precise instruments used to measure the mass of chemicals. Digital balances are commonly used for high accuracy.

Use: Accurate measurement of mass is crucial for quantitative chemical analysis, where the mass of reagents must be known to ensure proper stoichiometry in reactions.

k. Bunsen Burners

Description: A Bunsen burner is a small adjustable gas burner used for heating substances.

Use: It provides a controlled flame that can be used to heat glassware, liquids, or initiate chemical reactions that require heat.

l. Fume Hoods

Description: A fume hood is a ventilated enclosure in which hazardous chemicals can be handled safely.

Use: Fume hoods protect users from inhaling dangerous fumes, dust, or vapors during chemical reactions or while handling volatile or toxic chemicals.

m. Stirring Rods and Magnetic Stirrers

Description: Stirring rods are glass rods used to manually stir solutions. Magnetic stirrers use a magnetic bar placed in the solution and rotate it using a magnetic field to stir automatically.

Use: Stirring is essential for ensuring uniform mixing of chemicals in a solution, particularly during reactions or when dissolving a solid in a liquid.

2. Basic Laboratory Techniques

Mastering basic techniques is essential for achieving accurate results in the laboratory. Each technique requires careful attention to detail, ensuring that experiments are performed consistently and safely.

a. Measuring and Pouring Liquids

Techniques: Accurate measurements are crucial for experimental success. When using graduated cylinders or burettes, the liquid level should be read at the bottom of the meniscus (the curved surface of the liquid).

Tip: Ensure the equipment is on a flat surface and at eye level to avoid parallax error, which can lead to inaccurate readings.

b. Weighing Substances

Techniques: When using a balance, the substance should be weighed in a clean, dry container, such as a beaker or a watch glass. Balances should be calibrated frequently, and substances should never be placed directly on the balance pan to avoid contamination.

Tip: Always account for the mass of the container (tare the balance or subtract its weight from the final reading).

c. Titration

Techniques: In a titration, a solution of known concentration (titrant) is added to a solution of unknown concentration until the reaction reaches completion (usually indicated by a color change).

Steps: The burette is filled with the titrant, and small amounts are added to the solution being titrated while swirling. The endpoint is observed when the indicator changes color, signaling that the reaction is complete.

Tip: Record the volume of titrant added and repeat the titration several times to ensure accuracy.

d. Filtration

Techniques: Filtration is used to separate solid particles from a liquid. A filter paper is placed in a funnel, and the mixture is poured through. The solid is trapped by the filter paper, while the liquid passes through.

Tip: Ensure that the filter paper is properly fitted to avoid spillage, and use gravity or vacuum filtration depending on the requirements of the experiment.

e. Distillation

Techniques: Distillation is used to separate mixtures of liquids based on differences in boiling points. The liquid is heated until it boils, and the vapor is then condensed back into a liquid in a separate container.

Use: Commonly used for purifying liquids or separating volatile compounds.

Tip: Monitor the temperature carefully and ensure the condenser is working properly to avoid losing volatile components.

f. Crystallization

Techniques: Crystallization is a technique used to purify solid compounds. The impure solid is dissolved in a solvent at high temperature, and as the solution cools, pure crystals form.

Use: It is widely used for purifying chemicals or growing crystals for analysis.

Tip: Control the cooling process to allow slow crystal formation, leading to larger and purer crystals.

g. Heating Substances

Techniques: Substances are often heated to induce reactions or dissolve solutes. When using a Bunsen burner or hot plate, heat evenly and monitor the temperature to prevent accidents.

Tip: Use tongs or heat-resistant gloves to handle hot glassware, and avoid heating sealed containers to prevent explosions.

h. pH Measurement

Techniques: pH meters and indicators are used to measure the acidity or basicity of a solution. pH meters are electronic devices that provide an accurate numerical pH reading, while pH paper changes color based on the solution's pH.

Use: pH measurement is critical in acid-base chemistry, environmental testing, and biological applications.

Tip: Always calibrate the pH meter with standard buffer solutions before use for accurate readings.

i. Spectroscopy

Techniques: Spectroscopy involves analyzing the interaction between light and matter. Spectrophotometers measure how much light is absorbed by a substance, which can be used to determine the concentration of a chemical in solution.

Use: Widely used in analytical chemistry to identify compounds and study reaction kinetics.

Tip: Always clean the cuvettes (containers for solutions) to avoid contamination and ensure accurate results.

3. Safety in Using Laboratory Equipment

Proper handling of laboratory equipment is critical for maintaining a safe working environment. Accidents often occur when equipment is misused or when safety procedures are neglected. Here are a few general safety tips:

Check Equipment Before Use: Ensure that equipment, especially glassware, is free from cracks or defects before use. Damaged equipment can break under stress, leading to spills or injuries.

Use the Right Equipment: Always use the equipment designed for the task. For example, using a beaker to measure volume instead of a graduated cylinder can result in inaccuracies.

Handle with Care: Fragile equipment, especially glassware, should be handled carefully to avoid breakage. Use appropriate holders, clamps, or tongs when handling hot or delicate materials.

Follow Manufacturer's Instructions: For advanced equipment like balances, pH meters, or spectrophotometers, follow the operating manual to prevent damage and ensure accurate results.

Understanding the use and purpose of basic laboratory equipment and mastering fundamental techniques are foundational skills in chemistry. From beakers and test tubes to titrations and distillation, these tools and methods allow chemists to conduct experiments accurately and safely. Mastery of these elements is critical for success in both academic and industrial laboratory settings, as they form the core of all experimental procedures in chemistry. By adhering to proper protocols and safety guidelines, students and researchers can explore the fascinating world of chemistry while minimizing risks.

Chapter 2: The Basics of Matter

Definition and Classification of Matter

Understanding matter is fundamental to the study of chemistry. Matter is everything that occupies space and has mass, encompassing everything from the air we breathe to the water we drink and the complex compounds that make up living organisms. To better grasp the complexities of matter, it is essential to understand its definition and how it is classified into various categories.

1. Definition of Matter

Matter is defined as any substance that has mass and occupies space. This includes both tangible items we can see and touch, as well as substances that might be invisible or intangible in their natural state.

a. Mass

Definition: Mass is a measure of the amount of matter in an object or substance. It is usually measured in grams (g), kilograms (kg), or other units in the metric system.

Importance: Mass is a fundamental property of matter used to quantify how much of a substance is present and to perform calculations in chemical reactions and physical processes.

b. Volume

Definition: Volume is the amount of space that matter occupies. It is commonly measured in liters (L), milliliters (mL), or cubic meters (m^3).

Importance: Volume is used to understand the size of substances and is essential in determining concentrations and densities.

c. Substance

Definition: A substance is a form of matter with a specific composition and properties. Substances can be elements or compounds and are characterized by their chemical and physical properties.

2. Classification of Matter

Matter can be classified in various ways, but the primary classifications are based on its physical state and chemical composition. This helps in understanding its behavior, properties, and how it can be manipulated in chemical processes.

a. Physical State

Solid

Characteristics: Solids have a definite shape and volume. The particles in a solid are closely packed in a fixed arrangement, which gives solids their rigid structure. They have a low kinetic energy compared to liquids and gases.

Examples: Ice, metals, and wood.

Liquid

Characteristics: Liquids have a definite volume but take the shape of their container. The particles in a liquid are less closely packed than in a solid and can move past one another, allowing liquids to flow. They have higher kinetic energy than solids but lower than gases.

Examples: Water, alcohol, and oil.

Gas

Characteristics: Gases have neither a definite shape nor a definite volume. They expand to fill the shape and volume of their container. The particles in a gas are widely spaced and move freely, resulting in high kinetic energy.

Examples: Oxygen, carbon dioxide, and nitrogen.

Plasma

Characteristics: Plasma is an ionized gas with high energy levels, consisting of positively charged ions and free electrons. It does not have a definite shape or volume and is often found at extremely high temperatures.

Examples: Stars, lightning, and plasma TVs.

b. Chemical Composition

Elements

Definition: Elements are pure substances that cannot be broken down into simpler substances by chemical means. Each element consists of only one type of atom.

Examples: Hydrogen (H), Oxygen (O), and Carbon (C).

Periodic Table: Elements are organized in the Periodic Table, which groups them based on similar properties and atomic structure.

Compounds

Definition: Compounds are substances formed when two or more different elements are chemically bonded together. Compounds have a fixed composition and can be broken down into simpler substances through chemical reactions.

Examples: Water (H_2O), Carbon Dioxide (CO_2), and Sodium Chloride ($NaCl$).

Types of Bonds:

Ionic Bonds: Formed when electrons are transferred from one atom to another, creating positively and negatively charged ions.

Covalent Bonds: Formed when two atoms share one or more pairs of electrons.

Mixtures

Definition: Mixtures are combinations of two or more substances where each retains its own properties. Mixtures can be separated into their components by physical means.

Types of Mixtures:

Homogeneous Mixtures (Solutions): The components are uniformly distributed and indistinguishable.

Examples: Salt water, air, and alloys like bronze.

Heterogeneous Mixtures: The components are not uniformly distributed, and different substances can be distinguished.

Examples: Salad, sand and salt mixture, and oil and water.

3. Properties of Matter

Understanding the properties of matter is crucial in distinguishing between different substances and understanding their behavior in various conditions.

a. Physical Properties

- Definition: Physical properties can be observed or measured without changing the substance's chemical identity. - Examples: Color, melting and boiling points, density, and solubility.

b. Chemical Properties

- Definition: Chemical properties are observed when a substance undergoes a chemical change or reaction. These properties describe how a substance interacts with other substances. - Examples: Reactivity, flammability, and oxidation states.

4. Changes in Matter

Matter can undergo physical and chemical changes, which affect its properties and composition.

a. Physical Changes

- Definition: Physical changes affect only the appearance or state of matter but do not alter its chemical composition. - Examples: Melting ice, dissolving sugar in water, and chopping wood.

b. Chemical Changes

- Definition: Chemical changes result in the formation of new substances with different chemical compositions. This often involves a chemical reaction. - Examples: Burning wood, rusting of iron, and vinegar reacting with baking soda.

5. The Importance of Matter Classification

Classifying matter helps scientists and engineers understand its properties, predict how it will behave in various situations, and utilize it effectively in applications ranging from industrial processes to everyday products.

Understanding matter's types and behaviors is essential for conducting experiments, solving chemical problems, and applying chemistry to real-world challenges.

States of Matter: Solid, Liquid, and Gas

Understanding the states of matter is fundamental to chemistry as it helps explain how substances behave under various conditions. The three primary states of matter are solid, liquid, and gas. Each state has unique characteristics that influence how matter interacts with its surroundings.

1. Solid

In solids, particles are arranged in a fixed, orderly structure. This arrangement results in a solid having a definite shape and volume. The particles in a solid are tightly packed and vibrate around fixed positions, leading to the rigid nature of solids. This fixed arrangement prevents solids from changing shape or volume easily. Solids generally have higher densities compared to liquids and gases because their particles are closely packed. The density of a solid is relatively constant and does not vary significantly with temperature or pressure changes. Examples of solids include ice, metals like iron and aluminum, wood, and minerals such as quartz and diamond. Solids exhibit rigidity and resist deformation, maintaining their shape unless subjected to external forces. The temperature at which a solid melts into a liquid is called its melting point. This transition involves the absorption of heat, which provides enough energy for the particles to overcome the forces holding them together.

2. Liquid

Liquids have a definite volume but take the shape of their container. Unlike solids, liquids do not have a fixed shape and conform to the shape of the container they are in while maintaining a constant volume. In liquids, the particles are less tightly packed than in solids and can move past one another. This allows liquids to flow and adapt their shape to their container. Liquids have lower densities compared to solids but higher densities than gases. The density of a liquid can change with temperature and pressure, although it is generally less affected by pressure changes compared to gases. Common examples of liquids are water, milk, oil, and alcohol. Liquids are characterized by their fluidity, meaning they can flow and take the shape of their containers. The temperature at which a liquid turns into a gas is known as its boiling point. This phase transition requires heat to provide the necessary energy for the particles to escape the liquid phase.

3. Gas

Gases have neither a definite shape nor a definite volume. They expand to fill the shape and volume of their container due to the widely spaced particles that move freely in all directions. This results in gases having low density compared to solids and liquids. The density of a gas can vary significantly with temperature and pressure. Gas particles are far apart and move rapidly, resulting in the gas occupying the entire volume of its container and exerting pressure on its walls. Examples of gases include oxygen, carbon dioxide, nitrogen, and helium. Gases are highly compressible due to the large spaces between particles, allowing them to be compressed significantly. The pressure exerted by a gas is a result of collisions between gas particles and the walls of the container. This pressure is influenced by the temperature and volume of the gas, following the principles described by the ideal gas law.

4. Phase Transitions

Phase transitions describe the changes between different states of matter. Melting is the transition from solid to liquid, occurring when a solid absorbs heat and its particles gain enough energy to overcome their fixed positions. Freezing is the reverse transition, where a liquid loses heat, causing its particles to slow down and form a solid structure. Vaporization is the transition from liquid to gas, which includes evaporation (at the surface of the liquid) and boiling (throughout the liquid). Condensation is the transition from gas to liquid, occurring when a gas loses heat and its particles slow down to form a liquid. Sublimation is the transition from solid to gas without passing through the liquid state, such as with dry ice (solid carbon dioxide). Deposition is the reverse process, where a gas turns directly into a solid, such as frost forming on a cold surface.

Properties of Matter: Physical and Chemical

Understanding the properties of matter is essential for identifying substances, predicting their behavior, and manipulating them in various applications. Properties of matter are generally categorized into physical and chemical properties. Each type of property offers different insights into how matter behaves and interacts with other substances.

1. Physical Properties

Physical properties are characteristics that can be observed or measured without changing the substance's chemical identity. These properties describe the appearance and state of a substance and are often used to identify and characterize materials.

a. Extensive and Intensive Properties

Extensive Properties: These depend on the amount of matter present and change as the quantity of the substance changes. Examples include mass and volume. For instance, a larger sample of a substance will have more mass and occupy more volume than a smaller sample of the same substance.

Intensive Properties: These do not depend on the amount of matter present and remain constant regardless of the sample size. Examples include density, boiling point, and color. Intensive properties are particularly useful for identifying substances because they are characteristic of the material itself.

b. Common Physical Properties

Color: The visual appearance of a substance. Color can help in identifying substances and is often used in qualitative analysis.

Melting Point and Boiling Point: The temperatures at which a substance transitions from solid to liquid (melting point) and from liquid to gas (boiling point). These points are specific to each substance and are used to identify them and understand their behavior under different conditions.

Density: The mass per unit volume of a substance. It is calculated using the formula density = mass/volume. Density is a crucial property for identifying substances and understanding their buoyancy and interactions with other materials.

Solubility: The ability of a substance to dissolve in a solvent, such as water. Solubility is an important property in both industrial processes and biological systems.

Hardness: The resistance of a substance to deformation or scratching. Hardness is measured using scales such as the Mohs scale for minerals and is relevant in material science and engineering.

State of Matter: Solid, liquid, or gas. The state affects how substances interact with their environment and how they can be used in various applications.

c. Measurement and Observations

Physical properties can be observed directly or measured using various instruments and techniques. For example, a thermometer measures temperature to determine melting and boiling points, while a balance measures mass to help calculate density.

2. Chemical Properties

Chemical properties describe a substance's ability to undergo chemical changes or reactions, resulting in the formation of new substances with different chemical compositions. These properties are essential for understanding how substances interact and transform in chemical reactions.

a. Reactivity

Reactivity with Acids and Bases: Some substances react with acids or bases to form salts and other products. For example, metals like sodium react with acids to produce hydrogen gas and a salt.

Reactivity with Oxygen: Substances can react with oxygen to form oxides. For instance, iron reacts with oxygen to form rust (iron oxide).

b. Flammability

Definition: Flammability is the ability of a substance to burn or ignite, causing combustion. Highly flammable substances easily catch fire and burn, while less flammable substances require more energy to ignite.

Examples: Hydrocarbons like gasoline are highly flammable, while substances like water are not flammable.

c. Chemical Stability

Definition: Chemical stability refers to how resistant a substance is to chemical change or decomposition under normal conditions. Stable substances do not easily undergo chemical reactions, while unstable substances may decompose or react readily.

Examples: Noble gases like helium are chemically stable, while compounds like hydrogen peroxide are less stable and decompose over time.

d. Ability to Undergo Chemical Changes

Combustion: A chemical reaction where a substance reacts with oxygen to produce heat and light. Combustion typically results in the formation of carbon dioxide and water as products.

Oxidation-Reduction (Redox) Reactions: These involve the transfer of electrons between substances, leading to changes in oxidation states. For example, the rusting of iron is an oxidation reaction where iron loses electrons to oxygen.

Decomposition: The breakdown of a substance into simpler substances. For example, the decomposition of water into hydrogen and oxygen gas when electrolyzed.

3. Importance of Physical and Chemical Properties

Physical and chemical properties are crucial for various applications and fields, including:

Material Science: Understanding physical properties helps in selecting materials for construction, manufacturing, and technological applications.

Chemistry: Chemical properties guide reaction mechanisms, product formation, and the development of new compounds.

Environmental Science: Physical properties such as solubility and density are important in understanding environmental pollution and its effects.

Medicine: Chemical properties are essential in drug design, formulation, and the understanding of biochemical reactions in the body.

By studying and applying these properties, scientists and engineers can manipulate materials and reactions to achieve desired outcomes in both everyday life and specialized applications.

Changes in Matter: Physical and Chemical Changes

Matter undergoes various changes that can be classified into physical and chemical changes. Understanding these changes is crucial for analyzing and predicting the behavior of substances in different conditions.

1. Physical Changes

Physical changes involve alterations in a substance's physical properties without changing its chemical composition. These changes affect the form or appearance of a substance but do not result in the formation of a new substance.

a. Characteristics of Physical Changes

Reversible or Irreversible: Many physical changes are reversible, meaning the original state of the matter can be restored. For example, melting ice can be re-frozen to form solid ice. However, some physical changes are irreversible, like breaking a glass, where the original form cannot be perfectly restored, though no new substance is formed.

No New Substances Formed: In a physical change, the substance remains the same at the molecular level. There are no new substances created, just changes in state or appearance.

b. Examples of Physical Changes

Phase Changes: Changes between solid, liquid, and gas states are physical changes. Examples include melting (solid to liquid), freezing (liquid to solid), evaporation (liquid to gas), and condensation (gas to liquid). These changes involve energy transfer but do not alter the substance's chemical structure.

Dissolving: When a substance dissolves in a solvent, such as salt dissolving in water, it undergoes a physical change. The salt retains its chemical properties, and the solution can be separated back into its components by evaporation.

Breaking or Cutting: Breaking a piece of wood or cutting a sheet of paper into smaller pieces are physical changes. The material remains wood or paper, just in smaller or differently shaped pieces.

Mixing: Mixing two substances, like sand and salt, involves physical changes. The components retain their individual properties and can often be separated through physical means.

c. Measurement and Observation

Physical changes are often observed through changes in physical properties such as size, shape, state, or color. Measurement tools like thermometers (for temperature changes), scales (for mass), and graduated cylinders (for volume) can help quantify these changes.

2. Chemical Changes

Chemical changes involve transformations that result in the formation of new substances with different chemical compositions. These changes alter the chemical structure of the original substance, leading to new properties and products.

a. Characteristics of Chemical Changes

Formation of New Substances: A chemical change results in the formation of one or more new substances with different chemical properties. These new substances have different compositions from the original materials.

Energy Changes: Chemical changes often involve changes in energy, such as the release or absorption of heat. These energy changes can be observed as heat, light, or sound.

Irreversibility: Many chemical changes are irreversible under normal conditions, meaning the original substances cannot be easily recovered. For instance, burning wood results in ash and gases, which cannot be turned back into wood.

b. Examples of Chemical Changes

Combustion: Combustion is a chemical reaction where a substance reacts with oxygen, producing heat, light, and new products like carbon dioxide and water. For example, burning wood or gasoline involves combustion.

Oxidation: Oxidation involves the reaction of a substance with oxygen. A common example is rusting, where iron reacts with oxygen and moisture to form iron oxide (rust).

Acid-Base Reactions: These reactions occur when an acid reacts with a base to form water and a salt. For example, mixing hydrochloric acid (HCl) with sodium hydroxide ($NaOH$) results in water and sodium chloride (table salt).

Decomposition: Decomposition is a chemical change where a compound breaks down into simpler substances. For example, the decomposition of hydrogen peroxide (H_2O_2) into water (H_2O) and oxygen (O_2) when exposed to light or heat.

Fermentation: Fermentation is a biochemical process that converts sugars into alcohol and carbon dioxide with the help of yeast. An example is the fermentation of grapes to produce wine.

c. Indicators of Chemical Changes

Color Change: A change in color often indicates a chemical reaction. For example, the color change in a chemical indicator during a titration.

Gas Production: The formation of gas bubbles or fumes indicates a chemical change, such as the release of carbon dioxide during an acid-carbonate reaction.

Heat and Light: The release of heat or light during a reaction signifies a chemical change, as seen in combustion or fireworks.

Precipitate Formation: The formation of a solid from a solution during a reaction is called a precipitate. For example, mixing two clear solutions that form a cloudy, solid product.

3. Comparison of Physical and Chemical Changes

Physical Changes: Involve alterations in physical properties without changing the chemical composition. They are often reversible and do not produce new substances.

Chemical Changes: Involve changes in the chemical composition of substances, leading to the formation of new substances with different properties. They are usually irreversible and involve energy changes.

4. Importance of Understanding Changes in Matter

Understanding physical and chemical changes is vital for various applications in science and industry. This knowledge helps in:

Material Processing: Manufacturing processes often rely on both physical and chemical changes to produce desired materials and products.

Chemical Reactions: Predicting and controlling chemical reactions is essential for chemical synthesis, pharmaceuticals, and environmental applications.

Safety and Environmental Impact: Recognizing chemical changes helps in managing reactions and their byproducts, ensuring safety and minimizing environmental impact.

By studying these changes, scientists and engineers can manipulate matter to create new materials, design efficient processes, and address practical challenges.

Chapter 3: Atomic Structure

The Atom: Basic Concept

The atom is the fundamental unit of matter, serving as the basic building block of all substances. Understanding the atom's structure is essential for grasping concepts in chemistry and other sciences. Here's a detailed exploration of the atom's basic concept:

1. Definition and Importance

An atom is the smallest unit of an element that retains the chemical properties of that element. It consists of a central nucleus surrounded by a cloud of electrons. Atoms combine to form molecules, which make up all substances in the universe. Understanding atoms and their interactions is crucial for explaining chemical reactions, material properties, and many other phenomena in chemistry.

2. Historical Development

The concept of the atom has evolved over time, with significant contributions from several key scientists:

Democritus (c. 400 BCE): An ancient Greek philosopher who first proposed that matter is composed of indivisible particles called "atomos," meaning "uncuttable." Democritus suggested that different types of atoms made up different materials.

John Dalton (1803): Revived the atomic theory in the early 19th century. Dalton proposed that atoms are the fundamental building blocks of matter and that each element is made of identical atoms with unique properties. Dalton's theory also introduced the concept of atoms combining in fixed ratios to form compounds.

J.J. Thomson (1897): Discovered the electron using the cathode ray tube experiment. Thomson proposed the "plum pudding" model, where atoms are composed of a positively charged "pudding" with negatively charged electrons embedded within it.

Ernest Rutherford (1911): Conducted the gold foil experiment, which led to the discovery of the nucleus. Rutherford's experiment showed that atoms have a small, dense nucleus at their center, surrounded by electrons.

Niels Bohr (1913): Developed the Bohr model, which proposed that electrons orbit the nucleus in fixed energy levels or shells. This model explained the discrete lines observed in atomic spectra.

3. Atomic Structure

The basic structure of an atom consists of three primary subatomic particles:

Protons: Positively charged particles found in the nucleus. The number of protons in an atom determines the element's identity and is known as the atomic number. For example, an atom with one proton is hydrogen, and an atom with six protons is carbon.

Neutrons: Neutral particles (with no charge) also located in the nucleus. Neutrons contribute to the atom's mass but do not affect its charge. The number of neutrons, combined with protons, determines the atomic mass (or mass number) of an atom.

Electrons: Negatively charged particles orbiting the nucleus in electron shells or energy levels. Electrons are involved in chemical bonding and interactions. The number of electrons in a neutral atom equals the number of protons, balancing the atom's charge.

4. Atomic Models

Several atomic models have been proposed to explain the structure and behavior of atoms:

Dalton's Model: Proposed atoms as indivisible particles with specific weights and properties. This model viewed atoms as solid spheres.

Thomson's Model: Suggested that atoms are made up of a positively charged sphere with electrons embedded in it. This model could not explain the atom's structure observed in experiments.

Rutherford's Model: Introduced the idea of a dense, positively charged nucleus surrounded by electrons. This model highlighted the atom's central nucleus but did not explain electron arrangement or stability.

Bohr's Model: Proposed that electrons orbit the nucleus in fixed energy levels. Electrons can move between these levels by absorbing or emitting energy. This model successfully explained atomic spectra but was later refined with the development of quantum mechanics.

Quantum Mechanical Model: The current model, based on quantum mechanics, describes electrons as existing in probabilistic regions called orbitals, rather than fixed orbits. It uses principles such as wave-particle duality and Heisenberg's uncertainty principle to describe electron behavior.

5. Atomic Number and Mass Number

Atomic Number: The number of protons in an atom's nucleus. It defines the element and determines its position on the periodic table. For example, carbon has an atomic number of 6.

Mass Number: The total number of protons and neutrons in an atom's nucleus. It is used to calculate the atom's mass and differentiate between isotopes of the same element. For example, carbon-12 and carbon-14 are isotopes of carbon with mass numbers of 12 and 14, respectively.

6. Isotopes

Isotopes are variations of an element's atoms with the same number of protons but different numbers of neutrons. This results in different atomic masses but identical chemical properties. Isotopes are important in various applications, including dating methods (e.g., carbon dating) and medical diagnostics (e.g., radioactive tracers).

7. Electron Configuration

Electron configuration refers to the arrangement of electrons in an atom's electron shells or orbitals. This arrangement follows specific rules and principles:

Aufbau Principle: Electrons fill the lowest energy orbitals first before moving to higher energy levels.

Pauli Exclusion Principle: No two electrons in an atom can have the same set of quantum numbers. This principle explains the unique electron arrangement in orbitals.

Hund's Rule: Electrons occupy degenerate orbitals (orbitals of equal energy) singly before pairing up. This rule helps minimize electron-electron repulsion within orbitals.

8. Chemical Behavior and Bonding

The arrangement of electrons in an atom's outermost shell, known as valence electrons, determines its chemical behavior and bonding characteristics. Atoms interact to achieve stable electron configurations, often through forming chemical bonds such as ionic, covalent, or metallic bonds.

The atom is the fundamental unit of matter, with a complex structure consisting of a central nucleus surrounded by electrons. The historical development of atomic theory has led to a deeper understanding of atomic structure and behavior. Key concepts include the atomic number, mass number, isotopes, and electron configuration. Understanding these principles is crucial for studying chemistry, as they underpin the behavior and interaction of matter at the most fundamental level.

Subatomic Particles: Protons, Neutrons, and Electrons

The atom is the basic unit of matter, and its structure is defined by three fundamental subatomic particles: protons, neutrons, and electrons. Each of these particles plays a critical role in determining the properties and behavior of atoms.

1. Protons

a. Characteristics

Charge: Protons carry a positive electric charge, which is equal in magnitude but opposite in sign to the charge of an electron.

Mass: Protons have a mass of approximately 1 atomic mass unit (amu), which is about 1.67×10−271.67 \times 10^{-27}1.67×10−27 kilograms. This mass is roughly equivalent to that of neutrons and is much greater than the mass of electrons.

Location: Protons are located in the nucleus of the atom, which is the dense, central region of the atom.

b. Role in Atomic Structure

Atomic Number: The number of protons in an atom's nucleus defines the atomic number of an element. This number is unique to each element and determines its position on the periodic table. For instance, an atom with 6 protons is carbon.

Element Identity: The atomic number, or the number of protons, is fundamental to identifying the element. Changing the number of protons changes the element itself (e.g., changing a carbon atom to a nitrogen atom).

c. Examples

Hydrogen: The simplest atom, hydrogen, has 1 proton.

Oxygen: Oxygen has 8 protons.

2. Neutrons

a. Characteristics

Charge: Neutrons have no electric charge; they are neutral.

Mass: Neutrons have a mass of approximately 1 atomic mass unit (amu), similar to that of protons. This mass is slightly greater than the mass of an electron.

Location: Neutrons are found in the nucleus of the atom, along with protons.

b. Role in Atomic Structure

Atomic Mass: Neutrons contribute to the atomic mass of an atom. The mass number of an atom, which is the total number of protons and neutrons, is used to distinguish between different isotopes of an element.

Isotopes: Neutrons play a crucial role in the existence of isotopes. Isotopes of an element have the same number of protons but different numbers of neutrons. For example, carbon-12 and carbon-14 are isotopes of carbon with 6 protons and 6 or 8 neutrons, respectively.

c. Examples

Carbon Isotopes: Carbon-12 has 6 neutrons, while carbon-14 has 8 neutrons.

Uranium Isotopes: Uranium-238 has 146 neutrons, and uranium-235 has 143 neutrons.

3. Electrons

a. Characteristics

Charge: Electrons carry a negative electric charge, equal in magnitude but opposite in sign to the charge of a proton.

Mass: Electrons have a significantly smaller mass compared to protons and neutrons, approximately 9.11×10−319.11 \times 10^{-31}9.11×10−31 kilograms. Their mass is about 1/18361/18361/1836 of a proton's mass.

Location: Electrons orbit the nucleus in various energy levels or electron shells. They are not located in fixed paths but rather exist in probabilistic orbitals.

b. Role in Atomic Structure

Chemical Behavior: Electrons are primarily responsible for the chemical behavior of an atom. The arrangement of electrons in the outermost shell, known as valence electrons, determines how an atom interacts with other atoms and forms chemical bonds.

Electron Configuration: The distribution of electrons among the atom's orbitals is described by electron configuration. This configuration follows rules such as the Aufbau principle, Pauli exclusion principle, and Hund's rule to determine how electrons are arranged in atoms.

Charge Balance: In a neutral atom, the number of electrons equals the number of protons, resulting in a net charge of zero. However, atoms can gain or lose electrons to form ions, which have a net positive or negative charge.

c. Examples

Hydrogen: Hydrogen has 1 electron.

Neon: Neon has 10 electrons, arranged in 2 energy levels ($1s^2$ $2s^2$ $2p^6$).

4. Summary of Subatomic Particles

Protons: Positive charge, mass of 1 amu, located in the nucleus, determines the element's identity.

Neutrons: Neutral charge, mass of 1 amu, located in the nucleus, contributes to the atomic mass and isotope identity.

Electrons: Negative charge, negligible mass compared to protons and neutrons, orbit the nucleus, and determine chemical behavior.

5. Interactions Between Subatomic Particles

Electromagnetic Force: The electrostatic attraction between protons and electrons holds electrons in orbit around the nucleus. This force is responsible for the structure of the atom and the formation of chemical bonds.

Strong Nuclear Force: This force holds protons and neutrons together within the nucleus, overcoming the electrostatic repulsion between positively charged protons.

6. Significance in Chemistry

Element Identification: The number of protons (atomic number) identifies the element and determines its position on the periodic table.

Chemical Bonding: Electron configuration influences how atoms bond and interact. Understanding electron arrangement helps in predicting bonding patterns and reactivity.

Isotopic Analysis: Neutron number variation leads to isotopes, which have applications in dating, medical diagnostics, and nuclear energy.

Understanding protons, neutrons, and electrons provides a foundation for studying atomic interactions, chemical reactions, and material properties. Each subatomic particle has a specific role that contributes to the overall behavior and identity of the atom.

Atomic Models: From Dalton to Quantum Mechanics

The atomic model has undergone significant evolution since the early conceptions of matter. Each model has contributed to our current understanding of atomic structure, from John Dalton's solid sphere to the sophisticated quantum mechanical model. Here's an extensive look at the progression of atomic models:

1. Dalton's Atomic Model (Early 1800s)

a. Basic Concept

Solid Sphere Model: John Dalton proposed that atoms are indivisible, solid spheres. This model was based on Dalton's atomic theory, which was a major advance in understanding chemical reactions and matter.

Atomic Theory: Dalton's theory included the idea that all matter is composed of atoms, atoms of the same element are identical, and atoms combine in simple whole-number ratios to form compounds.

b. Contributions

Identification of Elements: Dalton's model provided a framework for understanding elements as pure substances composed of identical atoms.

Chemical Reactions: Dalton's model explained how atoms combine to form compounds and how they are rearranged in chemical reactions.

c. Limitations

Lack of Detail: Dalton's model could not explain the internal structure of the atom or the nature of the forces holding atoms together.

2. Thomson's Model (Late 1800s)

a. Discovery

Electron Discovery: J.J. Thomson discovered the electron in 1897 using the cathode ray tube experiment. This led to the development of the "plum pudding" model.

b. Plum Pudding Model

Positive Sphere with Embedded Electrons: Thomson's model proposed that the atom is a positively charged sphere with negatively charged electrons embedded within it, like "plums" in a "pudding."

Charge Neutrality: The positive charge was thought to balance the negative charge of the electrons, maintaining overall electrical neutrality.

c. Contributions

Electron Presence: Thomson's model was the first to introduce the concept of subatomic particles (electrons) and their negative charge.

d. Limitations

Inadequate Structure: The model could not explain the atomic spectrum or the arrangement of electrons in specific energy levels.

3. Rutherford's Model (Early 1900s)

a. Gold Foil Experiment

Discovery of the Nucleus: Ernest Rutherford conducted the gold foil experiment in 1911, which involved directing alpha particles at a thin gold foil. Most particles passed through, but some were deflected, indicating the presence of a dense, positively charged nucleus.

Nuclear Model: Rutherford proposed that the atom consists of a small, dense nucleus surrounded by orbiting electrons.

b. Contributions

Nucleus Concept: Rutherford's model introduced the concept of the nucleus as the central core of the atom, containing most of its mass and positive charge.

Electron Orbitals: Electrons were thought to orbit the nucleus, but the model could not explain why electrons did not spiral into the nucleus due to electromagnetic attraction.

c. Limitations

Electron Stability: The model could not account for the stability of electron orbits or explain atomic spectra.

4. Bohr's Model (1913)

a. Atomic Spectra Explanation

Quantized Energy Levels: Niels Bohr proposed that electrons orbit the nucleus in fixed, quantized energy levels or shells. Electrons could move between these levels by absorbing or emitting energy in discrete amounts.

Stable Orbits: Electrons in these quantized orbits do not radiate energy, thus maintaining stable orbits without spiraling into the nucleus.

b. Contributions

Hydrogen Spectrum: Bohr's model successfully explained the line spectra of hydrogen by correlating specific electron transitions with discrete energy levels.

Quantum Jumps: Electrons can jump between energy levels by absorbing or emitting photons, which correspond to the observed spectral lines.

c. Limitations

Complex Atoms: Bohr's model worked well for hydrogen but struggled to explain the spectra of more complex atoms or the fine details of atomic structure.

5. Quantum Mechanical Model (1920s-Present)

a. Development

Wave-Particle Duality: The quantum mechanical model, developed from the work of Schrödinger, Heisenberg, and others, incorporates the concept of wave-particle duality, where electrons exhibit both particle-like and wave-like properties.

Schrödinger's Equation: Schrödinger formulated an equation that describes how the quantum state of a physical system changes over time. The solutions to this equation give rise to atomic orbitals.

b. Electron Cloud Model

Probabilistic Orbitals: Instead of fixed orbits, electrons are described by probability distributions or orbitals. These orbitals represent regions where electrons are likely to be found.

Heisenberg Uncertainty Principle: The principle states that one cannot simultaneously know the exact position and momentum of an electron. This uncertainty is fundamental to the quantum mechanical description of atoms.

c. Contributions

Atomic Orbitals: The quantum mechanical model introduces the concept of orbitals (s, p, d, f) with different shapes and energy levels, explaining electron arrangement more accurately.

Pauli Exclusion Principle: This principle states that no two electrons in an atom can have the same set of quantum numbers, leading to the unique arrangement of electrons in orbitals.

d. Modern Applications

Molecular Orbital Theory: Extends the concept of atomic orbitals to molecules, explaining chemical bonding in terms of molecular orbitals.

Quantum Chemistry: Uses quantum mechanical principles to predict molecular behavior, reaction mechanisms, and spectroscopic properties.

6. Comparison of Models

Dalton's Model: Solid sphere, indivisible atoms.

Thomson's Model: Positively charged sphere with embedded electrons.

Rutherford's Model: Dense nucleus with orbiting electrons.

Bohr's Model: Quantized orbits for electrons.

Quantum Mechanical Model: Probabilistic orbitals, wave-particle duality.

The evolution of atomic models reflects advances in scientific understanding and technology. Dalton's model laid the foundation for atomic theory, Thomson's model introduced subatomic particles, Rutherford's model revealed the nucleus, Bohr's model provided insight into electron energy levels, and the quantum mechanical model offers a comprehensive description of atomic structure. Each model has contributed to our current understanding of the atom and its behavior, highlighting the dynamic nature of scientific inquiry and discovery.

Isotopes and Atomic Mass

1. Isotopes

a. Definition and Concept

Isotopes: Isotopes are variants of a chemical element that have the same number of protons but different numbers of neutrons in their atomic nuclei. This results in different atomic masses for these variants of the element.

Nuclear Stability: The difference in neutron number affects the stability of the atomic nucleus. Some isotopes are stable, while others are radioactive and decay over time.

b. Isotope Notation

Symbol Representation: Isotopes are often represented by the element's chemical symbol followed by the mass number (e.g., Carbon-12 or **C-12,** Carbon-14 or **C-14**).

General Formula: The notation is given as $^{A}_{Z}X$, where X is the chemical symbol of the element, A is the mass number (sum of protons and neutrons), and Z is the atomic number (number of protons).

c. Examples of Isotopes

Carbon Isotopes:

Carbon-12 (C-12): The most common isotope of carbon, with 6 protons and 6 neutrons.

Carbon-14 (C-14): A radioactive isotope used in radiocarbon dating, with 6 protons and 8 neutrons.

Hydrogen Isotopes:

Protium ($^{1}_{1}\mathrm{H}$): The most common isotope of hydrogen with 1 proton and 0 neutrons.

Deuterium ($^{2}_{1}\mathrm{H}$): An isotope of hydrogen with 1 proton and 1 neutron.

Tritium ($^{3}_{1}\mathrm{H}$): A radioactive isotope of hydrogen with 1 proton and 2 neutrons.

d. Applications of Isotopes

Radiocarbon Dating: Utilizes the decay of Carbon-14 to estimate the age of archaeological artifacts and geological samples.

Medical Imaging: Uses radioactive isotopes like Technetium-99m in imaging procedures to diagnose various medical conditions.

Environmental Tracking: Employs isotopic labeling to trace sources and pathways of pollutants in ecosystems.

2. Atomic Mass

a. Definition and Calculation

Atomic Mass: Atomic mass (also called atomic weight) is the weighted average mass of an atom of an element, accounting for the relative abundance of its isotopes.

Weighted Average: It is calculated by multiplying the mass of each isotope by its natural abundance (expressed as a fraction), then summing these values.

b. Formula

Calculation Formula: The atomic mass **A** is calculated using:

$$\mathrm{A} = \sum_i (\text{fractional abundance of isotope}_i \times \text{mass of isotope}_i)$$

c. Example

Chlorine: Chlorine has two main isotopes, Chlorine-35 and Chlorine-37. If the natural abundances are approximately 75% for Chlorine-35 and 25% for Chlorine-37, the atomic mass of chlorine is calculated as:

$$\text{Atomic Mass} = (0.75 \times 35) + (0.25 \times 37) = 26.25 + 9.25 = 35.50 \text{ amu}$$

d. Atomic Mass Units (amu)

Definition: The atomic mass unit (amu) is a standard unit of mass that quantifies the mass of atoms and molecules. One amu is defined as one twelfth the mass of a carbon-12 atom.

Usage: It is used to express atomic and molecular weights, making it easier to compare the relative masses of different atoms and molecules.

3. Isotopic Variation and Atomic Mass

a. Impact of Isotopes

Mass Variation: The presence of isotopes means that the atomic mass of an element is not always an integer. The atomic mass is an average value reflecting the contribution of each isotope's mass and abundance.

Elemental Analysis: In high-precision measurements, the variation in isotopic composition can affect the perceived mass of an element. Isotopic analysis is used to study variations in element concentrations and to understand natural processes.

b. Isotopic Abundance

Natural Abundance: The relative abundance of isotopes in nature can vary due to natural processes such as radioactive decay, nuclear reactions, and formation conditions. This abundance influences the calculated atomic mass of an element.

Variations: For example, Uranium has isotopes Uranium-235 and Uranium-238, with different natural abundances and uses in nuclear reactors and dating methods.

4. Summary

Isotopes: Variants of an element with the same number of protons but different numbers of neutrons, resulting in different atomic masses. They can be stable or radioactive and have various applications in science and industry.

Atomic Mass: The weighted average mass of an atom, calculated based on the relative abundance and mass of each isotope. It is measured in atomic mass units (amu) and provides a way to quantify the mass of atoms and molecules.

Understanding isotopes and atomic mass is crucial for various fields, including chemistry, physics, environmental science, and medicine. Isotopes offer insights into atomic structure and stability, while atomic mass allows for precise measurement and comparison of different elements and compounds.

Electron Configuration and Periodic Trends

1. Electron Configuration

a. Definition

Electron Configuration refers to the arrangement of electrons in the orbitals of an atom or molecule. It describes how electrons are distributed among energy levels, sublevels, and orbitals. Understanding electron configuration helps explain an atom's chemical properties, bonding behavior, and reactivity.

b. Structure of Electron Configuration

Energy Levels (Shells): Electrons occupy energy levels (shells) around the nucleus. These are labeled as **n = 1, 2, 3, ...**, where nnn represents the principal quantum number. The energy levels increase in energy as nnn increases.

Sublevels (Subshells): Each energy level contains one or more sublevels, which are labeled as **s, p, d,** and **f.** These sublevels have specific shapes and capacities for holding electrons:

s sublevel: Holds up to 2 electrons.

p sublevel: Holds up to 6 electrons.

d sublevel: Holds up to 10 electrons.

f sublevel: Holds up to 14 electrons.

Orbitals: Each sublevel contains orbitals, where electrons are likely to be found. An orbital can hold a maximum of two electrons with opposite spins.

c. Aufbau Principle

The **Aufbau Principle** (meaning "building up" in German) states that electrons fill orbitals starting with the lowest energy levels first before moving to higher ones. The typical order of filling follows:

$$1s, 2s, 2p, 3s, 3p, 4s, 3d, 4p, 5s, 4d, 5p, 6s, 4f, 5d, 6p, \text{ etc.}$$

d. Pauli Exclusion Principle

The **Pauli Exclusion Principle** states that no two electrons in an atom can have the same set of quantum numbers. This means that an orbital can hold only two electrons, and they must have opposite spins.

e. Hund's Rule

Hund's Rule states that when electrons occupy orbitals of the same sublevel, they will fill the orbitals singly before pairing up. This minimizes electron-electron repulsion and makes the atom more stable.

f. Electron Configuration Notation

Electron configuration is written by noting the energy levels, sublevels, and the number of electrons in each sublevel. For example:

Hydrogen: $1s^1$ (1 electron in the 1s1s1s sublevel).

Oxygen: $1s^2 2s^2 2p^4$ (2 electrons in the $1s$ sublevel, 2 in the $2s$, and 4 in the $2p$).

g. Orbital Diagrams

An **orbital diagram** is a visual representation of the arrangement of electrons in orbitals. Each orbital is represented by a box or line, and each electron is shown as an arrow. For example, the electron configuration of oxygen would be represented as:

g. Orbital Diagrams

An **orbital diagram** is a visual representation of the arrangement of electrons in orbitals. Each orbital is represented by a box or line, and each electron is shown as an arrow. For example, the electron configuration of oxygen would be represented as:

1s: ↑↓

2s: ↑↓

2p: ↑↓ ↑ ↑

h. Noble Gas Notation

A shorthand method for writing electron configurations is **noble gas notation**, where the core electrons (those in lower energy levels) are represented by the nearest noble gas. For example:

Calcium: $[Ar]4s^2$ (instead of writing the full configuration).

2. Periodic Trends

a. Definition

Periodic Trends refer to the recurring patterns in chemical properties that can be observed across periods (rows) and groups (columns) of the periodic table. These trends result from the arrangement of electrons and the structure of the atom.

b. Atomic Radius

Definition: The atomic radius is the distance from the center of the nucleus to the outermost electron.

Trend Across a Period: Atomic radius decreases across a period from left to right. This is because as protons are added to the nucleus, the increased nuclear charge pulls electrons closer to the nucleus.

Trend Down a Group: Atomic radius increases down a group. This is due to the addition of electron shells, which places outer electrons farther from the nucleus, despite the increased nuclear charge.

c. Ionization Energy

Definition: Ionization energy is the energy required to remove an electron from a gaseous atom or ion.

Trend Across a Period: Ionization energy increases across a period. As atomic radius decreases and nuclear charge increases, electrons are held more tightly by the nucleus, requiring more energy to remove one.

Trend Down a Group: Ionization energy decreases down a group. As atomic radius increases, the outermost electrons are farther from the nucleus and experience less attraction, making them easier to remove.

d. Electron Affinity

Definition: Electron affinity is the energy change that occurs when an atom gains an electron.

Trend Across a Period: Electron affinity generally becomes more negative (more energy is released) across a period. Atoms become more eager to gain electrons as they approach a full valence shell.

Trend Down a Group: Electron affinity becomes less negative down a group. As the atomic radius increases, the added electron is farther from the nucleus, experiencing less attraction and releasing less energy.

e. Electronegativity

Definition: Electronegativity is a measure of an atom's ability to attract electrons in a chemical bond.

Trend Across a Period: Electronegativity increases across a period. Atoms with smaller radii and higher nuclear charges more strongly attract bonding electrons.

Trend Down a Group: Electronegativity decreases down a group. Larger atoms with more electron shells have a weaker pull on bonding electrons.

f. Metallic and Non-Metallic Character

Metallic Character: The metallic character refers to how easily an atom loses electrons, a characteristic of metals.

Trend Across a Period: Metallic character decreases across a period, as atoms hold onto their electrons more tightly.

Trend Down a Group: Metallic character increases down a group, as larger atoms lose electrons more easily.

Non-Metallic Character: Non-metals tend to gain electrons in reactions.

Trend Across a Period: Non-metallic character increases across a period.

Trend Down a Group: Non-metallic character decreases down a group.

g. Summary of Key Periodic Trends

Property	Across a Period (left to right)	Down a Group (top to bottom)
Atomic Radius	Decreases	Increases
Ionization Energy	Increases	Decreases
Electron Affinity	Becomes more negative	Becomes less negative
Electronegativity	Increases	Decreases
Metallic Character	Decreases	Increases
Non-metallic Character	Increases	Decreases

3. Importance of Electron Configuration and Periodic Trends

a. Chemical Reactivity

The electron configuration of an atom determines how it interacts with other atoms, influencing bond formation, reactivity, and the types of compounds it can form. Atoms with incomplete valence shells are more reactive because they tend to gain, lose, or share electrons to achieve a stable electron configuration (often resembling the nearest noble gas).

b. Predicting Element Behavior

Periodic trends allow scientists to predict an element's chemical behavior based on its position in the periodic table. Elements in the same group tend to have similar properties because they have the same valence electron configuration.

c. Bonding and Compounds

Electronegativity differences between atoms influence the type of bond they form—whether it is ionic, covalent, or metallic. Understanding electron configuration helps explain how elements bond and why certain combinations of elements form particular compounds.

d. Industrial and Practical Applications

Knowledge of periodic trends and electron configurations is essential in materials science, pharmaceuticals, and environmental chemistry. For example, elements with high ionization energy are used in batteries, and isotopes with specific electron configurations are used in medical imaging and treatment.

Electron configuration and periodic trends are fundamental concepts in chemistry that explain the structure and behavior of atoms. Electron configuration provides insight into how electrons are arranged within an atom, while periodic trends reveal patterns in atomic properties that influence chemical reactivity, bonding, and element behavior across the periodic table. Together, these concepts offer a deeper understanding of the principles that govern atomic interactions and the chemical properties of elements.

Chapter 4: The Periodic Table

The Development of the Periodic Table

The periodic table is one of the most iconic and powerful tools in chemistry, organizing all known elements into a comprehensive and easy-to-understand framework. Its development reflects the evolution of scientific thought and discovery over centuries, culminating in the version we use today.

1. Early Attempts to Classify Elements

a. Alchemy and the Search for Order

In ancient times, before chemistry was established as a formal science, alchemists attempted to classify matter based on properties such as earth, water, air, and fire. Though not scientifically accurate, this early approach to understanding the nature of substances set the stage for later scientific advancements.

b. Antoine Lavoisier and the First Classification (1789)

The modern understanding of elements began with **Antoine Lavoisier**, often called the "Father of Modern Chemistry." In 1789, he compiled a list of 33 known elements, classifying them into four groups: gases, metals, nonmetals, and earths. Lavoisier's system, although rudimentary, was the first attempt at organizing elements based on observable physical and chemical properties. His work laid the groundwork for future efforts to understand the relationships between different elements.

2. Discovering Patterns and Properties

a. Johann Döbereiner and Triads (1829)

In 1829, **Johann Döbereiner** observed that certain elements with similar properties could be grouped into triads. For example, lithium (Li), sodium (Na), and potassium (K) formed one such triad, with the atomic weight of sodium being approximately the average of lithium and potassium. Although Döbereiner's classification was limited, it was an important step toward recognizing that there were patterns in elemental properties.

b. John Newlands and the Law of Octaves (1864)

In 1864, **John Newlands** proposed the **Law of Octaves**, suggesting that when elements were arranged by increasing atomic weight, every eighth element had similar properties, much like musical octaves. While this observation highlighted periodicity in element properties, Newlands' model was ridiculed at the time because it didn't work for all elements and was inconsistent with known chemical behavior. However, it provided a foundation for the idea of periodicity that would be key in later discoveries.

3. Dmitri Mendeleev and the First Periodic Table (1869)

a. Mendeleev's Breakthrough

Dmitri Mendeleev, a Russian chemist, is credited with creating the first widely recognized **Periodic Table of Elements** in 1869. Unlike his predecessors, Mendeleev organized the elements not only by increasing atomic weight but also by aligning them according to recurring chemical and physical properties. This approach allowed him to predict the existence of undiscovered elements and their properties with remarkable accuracy.

b. Gaps for Undiscovered Elements

Mendeleev's genius lay in his recognition that not all elements had been discovered. He left gaps in his table for elements that were yet to be identified, such as gallium (Ga), scandium (Sc), and germanium (Ge). When these elements were discovered later, their properties closely matched Mendeleev's predictions, providing strong evidence for the validity of his periodic table.

c. Strengths and Limitations

Mendeleev's table was groundbreaking because it successfully organized most of the known elements based on atomic weight and chemical properties. However, there were inconsistencies in his system, such as with **iodine (I)** and **tellurium (Te)**, where tellurium, though heavier, was placed before iodine because of its chemical behavior. Mendeleev prioritized chemical properties over atomic weight in these cases, which was difficult to explain at the time.

4. Henry Moseley and Atomic Number (1913)

a. Discovery of Atomic Number

In 1913, British physicist **Henry Moseley** made a key discovery that resolved many of the inconsistencies in Mendeleev's table. Using **X-ray spectroscopy**, Moseley determined that the elements are better organized by **atomic number** (the number of protons in an atom) rather than atomic weight. He demonstrated that the atomic number, not atomic mass, was the fundamental property that dictated the behavior of elements.

b. The Modern Periodic Law

With Moseley's discovery, the **Modern Periodic Law** was formulated: the properties of elements are periodic functions of their atomic numbers. This shifted the focus from atomic mass to atomic number as the basis for organizing the periodic table and eliminated the issues that had arisen in Mendeleev's version, such as the placement of iodine and tellurium.

5. The Role of Quantum Mechanics

a. Electron Configuration and Quantum Theory

The development of **quantum mechanics** in the early 20th century further refined the periodic table by providing a deeper understanding of the behavior of electrons in atoms. **Electron configuration**—the arrangement of electrons in shells and subshells around the nucleus—explained why elements in the same group of the periodic table have similar properties. It showed that elements are organized into blocks (s, p, d, f) based on their electron configurations, which corresponds to their chemical behavior.

b. Explanation of Periodicity

Quantum mechanics also explained the concept of **periodicity** in the properties of elements. As you move across a period, electrons fill the outermost orbitals, leading to changes in chemical reactivity. Moving down a group, elements have similar valence electron configurations, leading to similar chemical properties.

6. Expansion of the Periodic Table

a. The Discovery of Noble Gases (1894-1900)

The discovery of the **noble gases** (helium, neon, argon, krypton, xenon, and radon) between 1894 and 1900 added a new group to the periodic table. These elements were difficult to detect because of their inert nature but were eventually placed in a new group, Group 18, recognizing their unique lack of reactivity due to full valence electron shells.

b. Synthetic Elements and the Actinides

In the mid-20th century, scientists began synthesizing new elements that do not occur naturally, expanding the periodic table beyond uranium (element 92). These synthetic elements are part of the **actinide series** and include elements such as **plutonium (Pu)** and **americium (Am)**. These elements, particularly those with atomic numbers greater than 92, are typically radioactive and have short half-lives.

7. The Modern Periodic Table

a. Organization by Atomic Number

Today's periodic table is organized by **atomic number**, with elements arranged in rows (periods) and columns (groups) based on recurring chemical properties. The table includes **118 elements**, from hydrogen (H) to oganesson (Og), with periodicity in properties like atomic radius, ionization energy, and electronegativity evident across both rows and columns.

b. Element Blocks

The periodic table is divided into **blocks** based on electron configurations:

s-block: Groups 1 and 2 (alkali metals and alkaline earth metals).

p-block: Groups 13 to 18 (includes metals, metalloids, and nonmetals, such as halogens and noble gases).

d-block: Transition metals, located in the center of the table.

f-block: Lanthanides and actinides, often shown separately at the bottom.

c. Periodicity

The periodic table showcases periodic trends such as:

Atomic size decreases across a period and increases down a group.

Ionization energy increases across a period and decreases down a group.

Electronegativity increases across a period and decreases down a group.

8. Significance of the Periodic Table

The periodic table is not only a tool for organizing elements but also a predictive tool for understanding the behavior of elements and their compounds. It serves as the foundation for modern chemistry, allowing chemists to predict element reactivity, bond formation, and chemical reactions. The periodic table continues to evolve as new elements are synthesized, and our understanding of atomic structure deepens.

In summary, the development of the periodic table represents centuries of scientific progress, from the early attempts of classifying elements based on their physical and chemical properties to the modern arrangement based on atomic number and electron configuration. It remains one of the most essential tools in science, linking atomic structure to the chemical behavior of elements.

Understanding Periodic Trends: Groups and Periods

The periodic table is an essential tool for understanding the properties and behaviors of elements. Organized into **groups** (vertical columns) and **periods** (horizontal rows), the arrangement of the table reflects recurring trends, known as **periodic trends**, that help predict how elements will react chemically and physically. By examining the patterns within groups and periods, we can explain why certain elements behave similarly or differently.

1. Groups: Vertical Columns of the Periodic Table

a. Definition and General Characteristics of Groups

Groups are the vertical columns of the periodic table, and there are 18 groups in total. Elements in the same group share similar chemical properties because they have the same number of **valence electrons** (the electrons in the outermost shell). These valence electrons are primarily responsible for how an element participates in chemical reactions, such as forming bonds with other elements.

The elements in each group exhibit trends in **reactivity**, **electronegativity**, **atomic size**, and other properties as you move down the group. For example, the alkali metals in Group 1 (like lithium, sodium, and potassium) are all highly reactive, especially with water, due to their single valence electron.

b. Group Numbering and Element Categories

Groups in the periodic table are traditionally numbered 1–18. In older systems, they were often referred to as Group A or Group B elements, with the A groups being the main group elements and the B groups being the transition metals.

Elements in the same group tend to form similar compounds. For example, Group 17 elements, the **halogens** (such as fluorine, chlorine, and bromine), all form salts when they react with metals.

c. Trends within Groups

Atomic Radius: Increases down a group because each successive element has an additional electron shell, increasing the size of the atom.

Ionization Energy: Decreases down a group because the outermost electrons are farther from the nucleus and are more easily removed.

Electronegativity: Decreases down a group because the increased distance between the nucleus and the valence electrons reduces the attraction for bonding electrons.

Reactivity: In groups like the alkali metals (Group 1), reactivity increases as you move down the group. For halogens (Group 17), however, reactivity decreases as you move down.

2. Periods: Horizontal Rows of the Periodic Table

a. Definition and General Characteristics of Periods

Periods are the horizontal rows of the periodic table. There are seven periods, with Period 1 containing only two elements (hydrogen and helium), while Period 7 contains the largest number of elements, including the heavy, radioactive elements.

As you move from left to right across a period, the properties of elements change in a predictable way. This is due to the addition of protons to the nucleus and electrons to the outer energy levels, which affects atomic size, ionization energy, and electronegativity.

b. Trends across Periods

Atomic Radius: Decreases from left to right across a period because as protons are added to the nucleus, the increasing positive charge pulls the electrons closer to the nucleus, reducing the size of the atom.

Ionization Energy: Increases across a period because the increased nuclear charge holds the electrons more tightly, making it more difficult to remove an electron.

Electronegativity: Increases across a period because elements become more eager to attract electrons to achieve a full valence shell, especially as you approach the noble gases (Group 18).

Metallic Character: Decreases across a period. Elements on the left (such as the alkali metals) are highly metallic, meaning they are good conductors of heat and electricity. As you move to the right, elements become less metallic and more non-metallic (such as oxygen and fluorine), culminating in the noble gases, which are inert and non-metallic.

3. Detailed Examination of Periodic Trends

a. Atomic Radius Trend

Within a Group: Atomic radius increases as you move down a group. This is because each successive element has an additional electron shell, which increases the overall size of the atom. For example, in Group 1 (the alkali metals), lithium is the smallest, while cesium is much larger.

Within a Period: Atomic radius decreases from left to right across a period. As more protons are added to the nucleus, the increased nuclear charge pulls the electrons closer to the nucleus, making the atom smaller. For example, in Period 2, lithium has a much larger atomic radius than fluorine.

b. Ionization Energy Trend

Within a Group: Ionization energy decreases as you move down a group. Since the outer electrons are farther from the nucleus in larger atoms, they are less tightly bound and easier to remove. For instance, it is easier to remove an electron from cesium than from lithium in Group 1.

Within a Period: Ionization energy increases across a period. As atomic radius decreases and the nuclear charge increases, it becomes more difficult to remove an electron from an atom. For example, in Period 2, it is easier to remove an electron from lithium than from fluorine.

c. Electronegativity Trend

Within a Group: Electronegativity decreases down a group. Larger atoms have a weaker pull on shared electrons due to the increased distance between the nucleus and the valence electrons. In Group 17 (halogens), fluorine is the most electronegative, while iodine is less electronegative.

Within a Period: Electronegativity increases from left to right across a period. As atoms get closer to filling their valence shell, they have a stronger desire to attract electrons. For example, in Period 2, fluorine has the highest electronegativity, while lithium has a much lower value.

d. Metallic and Non-Metallic Character

Metallic Character: Metallic character increases down a group and decreases across a period. Elements that are more metallic are good conductors of electricity and heat, are malleable, and lose electrons easily to form positive ions. Alkali metals (Group 1) have a high metallic character, while non-metals like halogens and noble gases are less metallic.

Non-Metallic Character: Non-metallic character increases across a period and decreases down a group. Non-metals are poor conductors of heat and electricity, and they gain electrons easily to form negative ions. For instance, elements on the right side of the periodic table, such as oxygen and fluorine, are highly non-metallic.

4. Relationship Between Groups and Periods

The periodic table's organization into groups and periods is not arbitrary but rather reflects fundamental aspects of atomic structure and electron configuration. Elements in the same group have the same number of valence electrons, which gives them similar chemical properties. In contrast, elements in the same period have increasing numbers of protons and electrons, which leads to a regular change in properties across the row.

5. Special Groups and Their Periodic Trends

a. Alkali Metals (Group 1)

These elements are highly reactive, especially with water, and have a single valence electron. Reactivity increases as you move down the group, with cesium being more reactive than lithium.

b. Alkaline Earth Metals (Group 2)

These elements have two valence electrons and are less reactive than alkali metals but still highly reactive. Reactivity increases down the group, with barium being more reactive than magnesium.

c. Halogens (Group 17)

The halogens are highly reactive non-metals, with seven valence electrons. They are eager to gain one more electron to achieve a stable configuration. Reactivity decreases down the group, with fluorine being the most reactive.

d. Noble Gases (Group 18)

These elements have full valence electron shells, making them highly stable and largely unreactive. Noble gases are often used in lighting and other applications where non-reactivity is important.

6. Importance of Periodic Trends in Chemistry

Understanding periodic trends is crucial for predicting the behavior of elements during chemical reactions. These trends help chemists anticipate how an element will bond, what kinds of compounds it will form, and how it will interact with other elements. The periodic table is a predictive tool, allowing scientists to deduce unknown properties of elements and predict new ones based on their position in the table.

Conclusion

Periodic trends in groups and periods form the foundation of modern chemistry. By understanding how properties like atomic radius, ionization energy, and electronegativity change across groups and periods, chemists can predict the behavior of elements and compounds in a wide variety of situations. This understanding not only deepens our knowledge of chemical reactions but also enables the discovery of new elements and materials with useful properties.

Classification of Elements: Metals, Nonmetals, and Metalloids

The periodic table organizes elements not only by their atomic number and properties but also by their classification into three broad categories: **metals**, **nonmetals**, and **metalloids**. Each class has distinctive physical and chemical properties, which are reflected in their position on the periodic table. Understanding these classifications is fundamental to grasping how elements behave in chemical reactions and how they are used in various practical applications.

1. Metals

Metals are the most abundant category of elements, found primarily on the left and center of the periodic table. They include familiar elements like iron, copper, gold, and aluminum.

a. Physical Properties of Metals

Luster (Shiny Appearance): Most metals have a reflective, shiny surface when polished.

Malleability and Ductility: Metals can be hammered into thin sheets (malleability) or drawn into wires (ductility) without breaking.

Conductivity: Metals are excellent conductors of heat and electricity due to the free movement of electrons within their structure.

High Melting and Boiling Points: Metals generally have high melting and boiling points, which makes them stable at room temperature (with a few exceptions like mercury, which is liquid at room temperature).

High Density: Metals are typically dense and heavy because of closely packed atoms in a solid structure.

b. Chemical Properties of Metals

Electropositivity: Metals tend to lose electrons easily during chemical reactions, forming positive ions (cations). This ability to donate electrons makes them good reducing agents.

Reactivity with Nonmetals: Metals readily react with nonmetals, especially oxygen and halogens, to form ionic compounds such as metal oxides (e.g., rusting of iron) and metal halides (e.g., sodium chloride).

Corrosion: Many metals corrode when exposed to oxygen and water. For instance, iron rusts when exposed to moist air, although some metals, like aluminum, form a protective oxide layer that prevents further corrosion.

c. Location of Metals on the Periodic Table

Alkali Metals (Group 1): Highly reactive metals like lithium, sodium, and potassium, which react vigorously with water.

Alkaline Earth Metals (Group 2): Slightly less reactive than alkali metals but still reactive, including elements like magnesium and calcium.

Transition Metals (Groups 3-12): Metals like iron, copper, and gold that exhibit various oxidation states and are often used in construction, electronics, and jewelry.

Other Metals: Elements such as aluminum, tin, and lead found in groups 13-16 but still exhibit typical metallic properties.

2. Nonmetals

Nonmetals, found on the upper right side of the periodic table, exhibit properties that are the opposite of metals. They include elements like oxygen, carbon, sulfur, and nitrogen.

a. Physical Properties of Nonmetals

Lack of Luster: Nonmetals do not have the shiny appearance that metals do; they tend to have dull surfaces.

Brittle: In solid form, nonmetals are typically brittle and will shatter if struck.

Poor Conductors: Nonmetals are poor conductors of heat and electricity, making them good insulators. For example, sulfur and phosphorus are non-conductive.

Lower Melting and Boiling Points: Nonmetals generally have lower melting and boiling points than metals. Many nonmetals are gases at room temperature, such as oxygen and nitrogen, while some exist as solids, like carbon.

b. Chemical Properties of Nonmetals

Electronegativity: Nonmetals tend to gain electrons during chemical reactions, forming negative ions (anions). They have high electronegativity, meaning they attract electrons strongly in chemical bonds.

Reactivity with Metals and Other Nonmetals: Nonmetals react with metals to form ionic compounds (e.g., sodium chloride). When reacting with other nonmetals, they form covalent compounds (e.g., carbon dioxide).

Oxidizing Agents: Nonmetals often act as oxidizing agents, meaning they accept electrons from other elements during chemical reactions. Oxygen, for instance, is a powerful oxidizing agent.

c. Location of Nonmetals on the Periodic Table

Nonmetals are located on the upper right side of the periodic table. The most notable nonmetals are found in:

Group 17 (Halogens): Elements like fluorine, chlorine, and iodine, which are highly reactive nonmetals, especially with alkali and alkaline earth metals.

Group 16 (Oxygen Group): Elements like oxygen and sulfur.

Group 18 (Noble Gases): These are inert nonmetals like helium, neon, and argon, which do not readily form compounds due to their full valence electron shells.

3. Metalloids

Metalloids, also known as **semimetals**, possess properties that are intermediate between metals and nonmetals. They are located along the "stair-step" line that separates metals and nonmetals on the periodic table, and they include elements like silicon, germanium, and arsenic.

a. Physical Properties of Metalloids

Appearance: Metalloids often have a metallic luster but are more brittle than metals.

Semiconductors: Metalloids are semiconductors, meaning they can conduct electricity under certain conditions but not as effectively as metals. This property makes them invaluable in the electronics industry.

Moderate Density and Melting Points: The melting points and densities of metalloids are generally between those of metals and nonmetals.

b. Chemical Properties of Metalloids

Variable Reactivity: Metalloids can either lose or gain electrons during chemical reactions, depending on the elements they are reacting with. For example, silicon typically behaves as a nonmetal, but under certain conditions, it can act like a metal.

Amphoteric Behavior: Metalloids often exhibit both acidic and basic properties, depending on their reaction environment. This versatility allows them to form both covalent and ionic compounds.

c. Location of Metalloids on the Periodic Table

Metalloids form a diagonal boundary between the metals and nonmetals, creating a step-like division. The major metalloids include:

- **Boron (B)**
- **Silicon (Si)**
- **Germanium (Ge)**
- **Arsenic (As)**
- **Antimony (Sb)**
- **Tellurium (Te)**

4. Comparing Metals, Nonmetals, and Metalloids

Property	Metals	Nonmetals	Metalloids
State at Room Temp	Solid (except mercury)	Solid, liquid, or gas	Solid
Luster	Shiny	Dull	Can be shiny or dull
Conductivity	Good conductors	Poor conductors (insulators)	Semiconductors
Malleability	Malleable	Brittle	Brittle
Electron Behavior	Lose electrons (form cations)	Gain electrons (form anions)	Can lose or gain electrons
Bond Formation	Form metallic and ionic bonds	Form covalent and ionic bonds	Form covalent bonds

5. Significance of Element Classification

Understanding the classification of elements as metals, nonmetals, or metalloids is crucial in predicting their behavior in chemical reactions and their practical applications in industry and daily life.

a. Industrial Applications

Metals: Widely used in construction, transportation, and manufacturing due to their strength, conductivity, and malleability. For example, steel is used in buildings, and copper is used in electrical wiring.

Nonmetals: Essential in chemical industries, medicine, and agriculture. For example, oxygen is critical for respiration and combustion processes, while nitrogen is used to produce fertilizers.

Metalloids: Crucial in the electronics industry due to their semiconductor properties. Silicon, for instance, is the backbone of modern computer chips and solar panels.

b. Biological Significance

Metals: Certain metals like iron (in hemoglobin) and calcium (in bones) are vital for biological processes.

Nonmetals: Nonmetals like carbon, hydrogen, oxygen, and nitrogen are the building blocks of life, forming carbohydrates, proteins, lipids, and nucleic acids.

Metalloids: Elements like boron and silicon play roles in plant and animal life, though they are less prominent than metals and nonmetals.

The classification of elements into metals, nonmetals, and metalloids provides a framework for understanding the periodic table's organization and the fundamental differences in element behavior. Metals are generally good conductors and malleable, nonmetals are insulators and brittle, while metalloids bridge the gap with mixed properties. These classifications not only predict how elements interact in chemical reactions but also have practical applications in industries like electronics, construction, and healthcare.

The Role of the Periodic Table in Chemistry

The periodic table is often referred to as the **"blueprint of chemistry"** because it organizes all known chemical elements in a systematic manner, allowing chemists to predict and understand the behavior of elements based on their position. Its importance extends beyond just being a reference tool—it is a **framework** that explains chemical reactions, properties, and the relationships between elements. The periodic table plays an indispensable role in chemistry by providing insight into **element classification, trends, and bonding**.

1. Organizing and Classifying Elements

The periodic table arranges elements by increasing atomic number (the number of protons in an atom's nucleus), which creates a logical order based on the element's structure and properties. This organization reveals patterns in chemical behavior that are critical to understanding how elements interact.

a. Groups (Columns)

Vertical columns in the periodic table are known as **groups** or **families**. Elements within the same group share similar chemical properties because they have the same number of **valence electrons** (the electrons in the outermost shell of an atom). These valence electrons determine how an element bonds with others.

Example: All elements in **Group 1 (Alkali metals)**, such as lithium, sodium, and potassium, have one valence electron, making them highly reactive and prone to losing that electron to form positive ions.

b. Periods (Rows)

Horizontal rows are called **periods**. As you move from left to right across a period, the atomic number increases, and the properties of the elements change in a predictable manner. Elements in the same period have the same number of **electron shells**, but different numbers of valence electrons.

Example: In **Period 2**, elements like lithium (Li), beryllium (Be), and fluorine (F) all have two electron shells, but their chemical behavior changes because they have different numbers of electrons in their outer shell.

2. Predicting Chemical Behavior

The periodic table allows chemists to predict how elements will react chemically based on their position. Elements in the same group, for instance, exhibit similar chemical reactivity, while trends across periods reveal how atomic structure influences reactivity, electronegativity, and bonding capabilities.

a. Metals, Nonmetals, and Metalloids The periodic table helps to **categorize elements** into **metals, nonmetals, and metalloids** based on their chemical and physical properties. This categorization enables chemists to predict behaviors like bonding tendencies:

Metals, found on the left side, are typically electron donors, forming positive ions in chemical reactions.

Nonmetals, located on the right, tend to accept electrons, forming negative ions or sharing electrons in covalent bonds.

Metalloids, situated along the dividing line between metals and nonmetals, display mixed properties, such as semiconductivity.

b. Element Reactivity The periodic table is key to understanding **element reactivity**:

Highly reactive metals like those in Group 1 (alkali metals) lose electrons easily, reacting vigorously with water and halogens.

Halogens (Group 17) are highly reactive nonmetals because they need only one electron to complete their valence shell.

Noble gases (Group 18) are largely unreactive due to their stable, full electron configurations.

3. Periodic Trends and Their Importance

The periodic table is organized so that certain **trends** become apparent. These trends allow chemists to infer the properties of unknown or less familiar elements based on their location in the table. The key trends include **atomic radius, ionization energy, electronegativity**, and **electron affinity**.

a. Atomic Radius

The **atomic radius** refers to the size of an atom. As you move down a group, the atomic radius increases because additional electron shells are added. However, as you move across a period from left to right, the atomic radius decreases, as the increasing number of protons pulls the electrons closer to the nucleus.

Example: In Group 1, lithium (Li) is smaller than potassium (K) because potassium has more electron shells.

b. Ionization Energy

Ionization energy is the energy required to remove an electron from an atom. It increases across a period from left to right due to stronger attraction between the nucleus and the electrons. Conversely, it decreases down a group as the distance between the nucleus and the valence electrons increases, making it easier to remove an electron.

Example: Sodium (Na) has lower ionization energy than chlorine (Cl) because it is easier to remove an electron from sodium.

c. Electronegativity

Electronegativity is the ability of an atom to attract electrons in a chemical bond. It increases across a period because atoms become more eager to fill their valence shell, and it decreases down a group because the added electron shells reduce the effective nuclear charge on the outer electrons.

Example: Fluorine (F) is the most electronegative element, meaning it strongly attracts electrons in a bond.

d. Electron Affinity

Electron affinity is the energy change that occurs when an atom gains an electron. Generally, nonmetals have high electron affinity, as they are close to filling their valence shells, while metals have lower electron affinity since they prefer to lose electrons.

Example: Halogens like chlorine (Cl) have high electron affinity because gaining an electron allows them to achieve a stable configuration.

4. Chemical Bonding and the Periodic Table

The periodic table is a crucial tool for understanding how atoms bond to form compounds. The type of bond—whether **ionic, covalent, or metallic**—can often be predicted based on the elements involved and their positions in the table.

a. Ionic Bonding

Ionic bonds typically form between metals and nonmetals. Metals, especially those in Groups 1 and 2, lose electrons to become cations, while nonmetals (especially Group 17 elements) gain those electrons to form anions.

Example: Sodium (Na) donates an electron to chlorine (Cl), forming the ionic compound sodium chloride ($NaCl$).

b. Covalent Bonding

Covalent bonds occur when atoms, usually nonmetals, share electrons. Elements in Groups 14 to 16 tend to form covalent bonds because they prefer to share electrons to fill their valence shells.

Example: Oxygen (O) forms covalent bonds with hydrogen (H) in water (H_2O), sharing electrons to achieve stability.

c. Metallic Bonding

Metallic bonds are formed between metal atoms. In these bonds, electrons are delocalized and free to move, which explains why metals conduct electricity and are malleable.

Example: In solid copper (Cu), metallic bonding allows the metal to conduct electricity efficiently.

5. The Periodic Table as a Predictive Tool

One of the most significant roles of the periodic table in chemistry is its ability to **predict the properties of elements** and **newly discovered elements**. The arrangement of elements into periods and groups allows scientists to forecast how an element will behave even if they haven't fully explored it experimentally.

a. Discovery of New Elements When scientists discover new elements, they can place them into the periodic table based on their atomic number. From their position, chemists can predict the element's properties, reactivity, and potential uses before extensive testing.

Example: The discovery of elements like **oganesson (Og)**, the heaviest known element, fits into Group 18 (the noble gases), and predictions about its properties stem from its position.

b. Designing New Compounds Chemists use the periodic table to design **new compounds** by predicting how elements will react with one another. This predictive power has led to advancements in fields such as pharmaceuticals, materials science, and environmental chemistry.

6. Applications in Industry, Medicine, and Technology

The periodic table serves as a vital tool across industries, guiding research and development in chemistry-based fields. Its utility extends far beyond the laboratory, influencing medicine, energy, and technology.

a. Medicine Pharmaceutical chemists rely on the periodic table to design drugs by understanding how different elements interact with biological systems. Elements such as **platinum** are used in chemotherapy, and compounds containing metals like **iron** or **magnesium** are essential for various medical treatments.

b. Technology In electronics and materials science, the periodic table helps identify elements suitable for developing **semiconductors**, **batteries**, and **solar panels**. **Silicon**, a metalloid, is essential for semiconductor technology, while **lithium** is crucial for energy storage in rechargeable batteries.

c. Environmental Chemistry The periodic table also plays a role in developing solutions to **environmental challenges**. By understanding how elements interact with the environment, chemists can create materials for pollution control, such as **catalysts** that reduce harmful emissions or filters that remove toxic metals from water.

The periodic table is more than a chart of elements; it is the foundational tool of modern chemistry. It provides a comprehensive framework for understanding the properties and behaviors of elements, predicting chemical reactions, and guiding scientific discovery. Its role in organizing elements by atomic number and revealing trends across groups and periods allows chemists to predict how elements will behave in chemical reactions, making it indispensable in education, research, and industry. Through its detailed organization and predictive power, the periodic table remains a central pillar of chemistry that continues to drive innovation across various scientific fields.

Chapter 5: Chemical Bonding

Types of Chemical Bonds

Chemical bonds are the forces that hold atoms together to form molecules and compounds. These bonds result from interactions between atoms' electrons, particularly those in the outermost shell, called **valence electrons**. The nature of the bond between atoms determines the physical and chemical properties of the resulting substance. There are three main types of chemical bonds: **ionic bonds, covalent bonds**, and **metallic bonds**. Each bond type arises from a different way that atoms share or transfer electrons to achieve stability.

1. Ionic Bonds

An **ionic bond** is formed through the **transfer of electrons** between atoms, typically between a metal and a nonmetal. This type of bonding occurs when one atom (usually a metal) donates one or more of its valence electrons to another atom (usually a nonmetal), leading to the formation of oppositely charged ions. These ions are then held together by strong electrostatic forces.

a. Formation of Ionic Bonds

Metals, which are found on the left side of the periodic table, tend to have low ionization energies, meaning they can lose electrons easily. When a metal atom loses electrons, it becomes a positively charged ion, or **cation**.

Nonmetals, located on the right side of the periodic table, have high electron affinity and gain electrons to become negatively charged ions, or **anions**.

The resulting cations and anions attract each other due to their opposite charges, forming a strong ionic bond.

Example: Sodium Chloride (NaCl) In the case of **sodium chloride (table salt)**, a sodium atom (Na) loses one electron to form a Na^+ ion, while a chlorine atom (Cl) gains that electron to form a Cl^- ion. These oppositely charged ions attract each other, resulting in the formation of NaCl.

b. Properties of Ionic Compounds Ionic compounds tend to have distinct physical and chemical properties:

High melting and boiling points: Ionic bonds are strong, so it takes a significant amount of energy to break them.

Electrical conductivity: Ionic compounds conduct electricity when dissolved in water or melted, as the ions are free to move.

Solubility in water: Many ionic compounds dissolve easily in water due to the interaction between the ions and the polar water molecules.

2. Covalent Bonds

A **covalent bond** forms when two atoms, usually nonmetals, **share electrons** to achieve stability. Unlike ionic bonds, where electrons are transferred, covalent bonding involves a mutual sharing of electrons, allowing both atoms to achieve a full outer electron shell. Covalent bonds can be single, double, or triple, depending on how many pairs of electrons are shared between the atoms.

a. Formation of Covalent Bonds

Covalent bonds usually form between atoms with similar electronegativities, meaning neither atom has a strong tendency to gain or lose electrons.

Atoms share pairs of electrons so that each can attain a stable electron configuration, similar to that of a noble gas.

A **single covalent bond** involves the sharing of one pair of electrons, a **double bond** involves two pairs, and a **triple bond** involves three pairs of electrons shared between atoms.

Example: Water (H_2O) In **water**, each hydrogen atom shares one electron with the oxygen atom, forming two single covalent bonds. Oxygen has a higher electronegativity than hydrogen, so the electrons are not shared equally, leading to a polar covalent bond.

b. Types of Covalent Bonds

Nonpolar covalent bonds: Electrons are shared equally between atoms, resulting in no partial charges. This typically occurs when atoms of the same element bond together (e.g., O_2 or H_2).

Polar covalent bonds: Electrons are shared unequally, creating partial positive and negative charges on the bonded atoms. This happens when atoms with different electronegativities bond, as seen in water (H_2O).

c. Properties of Covalent Compounds Covalent compounds have different properties compared to ionic compounds:

Lower melting and boiling points: Covalent bonds are generally weaker than ionic bonds, so less energy is needed to break them.

Poor conductivity: Covalent compounds do not conduct electricity in water because they do not break into ions.

Solubility: Covalent compounds may or may not dissolve in water, depending on whether they are polar or nonpolar. Polar covalent compounds (like sugar) tend to dissolve, while nonpolar compounds (like oil) do not.

3. Metallic Bonds

A **metallic bond** occurs between metal atoms, where **valence electrons are shared collectively** by a lattice of atoms. In metallic bonding, the electrons are not bound to any one atom but move freely throughout the entire structure, creating what is often referred to as a "sea of electrons." This type of bonding explains the unique properties of metals.

a. Formation of Metallic Bonds

In metals, atoms are arranged in a lattice structure where valence electrons are free to move between atoms.

The positively charged metal ions are held together by the shared "sea of electrons." This free movement of electrons gives metals their ability to conduct electricity and heat.

Metallic bonds are non-directional, meaning that the bonding strength is spread out over the entire metal structure, allowing for malleability and ductility.

Example: Copper (Cu) In **copper**, the metal atoms share their valence electrons, which are free to move through the copper lattice, making copper an excellent conductor of electricity.

b. Properties of Metallic Bonds Metals exhibit several unique properties due to metallic bonding:

Electrical conductivity: The free movement of electrons in the "sea of electrons" allows metals to conduct electricity.

Thermal conductivity: The delocalized electrons also transfer heat efficiently, making metals good conductors of heat.

Malleability and ductility: Because metallic bonds are not rigid, metals can be hammered into sheets (malleable) or drawn into wires (ductile) without breaking.

Shiny appearance: Metals tend to be shiny due to the way their electrons reflect light.

4. Comparison of Bond Types

Understanding the differences between ionic, covalent, and metallic bonds helps explain why materials exhibit different properties:

Property	Ionic Bonds	Covalent Bonds	Metallic Bonds
Formation	Electron transfer between atoms	Electron sharing between atoms	Electron pooling among atoms
Bond Strength	Strong due to electrostatic forces	Varies: can be weak or strong	Strong but flexible
Conductivity	Conductive in liquid/solution	Poor conductor (except in solutions)	Highly conductive
Melting/Boiling Point	High	Low to moderate	High
Malleability	Brittle	Brittle	Malleable and ductile
Examples	NaCl, MgO	H_2O, CO_2	Copper, Aluminum

5. Other Types of Bonds

In addition to the three main types of bonds, there are several other important bonding interactions that play a crucial role in chemical reactions and the properties of substances:

a. Hydrogen Bonds A **hydrogen bond** is a weak interaction that occurs between a hydrogen atom covalently bonded to a highly electronegative atom (such as nitrogen, oxygen, or fluorine) and another electronegative atom. Hydrogen bonding is essential for the properties of water and plays a critical role in biological molecules like DNA.

Example: Water (H_2O) In water, hydrogen bonds form between the hydrogen of one water molecule and the oxygen of another, contributing to water's high boiling point and surface tension.

b. Van der Waals Forces These are weak, temporary attractions between molecules or atoms caused by momentary fluctuations in electron distribution. Van der Waals forces include **London dispersion forces** (present in all molecules) and **dipole-dipole interactions** (which occur in polar molecules).

Chemical bonding is fundamental to understanding the structure and behavior of substances. Ionic, covalent, and metallic bonds represent the primary ways in which atoms combine to form compounds and materials. The type of bonding dictates the properties of substances, influencing their melting points, conductivity, reactivity, and physical strength. Mastery of these bonding concepts is crucial for deeper exploration into more complex chemical reactions, molecular structures, and the functionality of materials in various applications.

Bond Formation and Properties

Chemical bonding is the process by which atoms combine to form more complex substances such as molecules, crystals, and compounds. The way atoms bond together determines the properties of the resulting substance, from its stability and structure to its reactivity and behavior under different conditions. At the heart of bond formation is the tendency of atoms to achieve stability, often by filling or emptying their outer electron shells. This desire for stability drives atoms to either transfer, share, or pool their electrons, leading to the formation of different types of bonds.

Understanding bond formation and the properties associated with each bond type is fundamental to mastering chemistry and predicting the behavior of compounds. In this section, we will explore the various processes involved in bond formation and their associated properties.

1. Why Bonds Form: The Octet Rule

Atoms form bonds to achieve a **stable electron configuration**, which usually means having a full valence shell of electrons. The **octet rule** is the guiding principle behind bond formation in many elements, especially for main-group elements. According to the octet rule, atoms are most stable when they have eight electrons in their outermost shell (or two for hydrogen and helium).

In ionic bonds, atoms either gain or lose electrons to fill their outer shell.

In covalent bonds, atoms share electrons to fulfill the octet rule.

In metallic bonds, electrons are pooled and shared freely among all atoms in a metallic lattice, achieving stability.

While the octet rule applies to many elements, there are exceptions, especially among transition metals and molecules with fewer or more than eight electrons around an atom.

2. Ionic Bond Formation

An **ionic bond** forms through the **transfer of electrons** between atoms. This typically occurs between a **metal and a nonmetal**, where one atom (usually a metal) donates one or more electrons, and the other atom (usually a nonmetal) accepts them.

Cation Formation (Metal): Metals tend to have low ionization energy, meaning they lose electrons easily. When a metal atom loses one or more electrons, it becomes a **positively charged ion**, or **cation**.

Example: Sodium (Na), a metal, has one electron in its outer shell. To achieve stability, it loses this electron to become Na^+.

Anion Formation (Nonmetal): Nonmetals have high electron affinity, meaning they readily accept electrons. When a nonmetal gains one or more electrons, it becomes a **negatively charged ion**, or **anion**.

Example: Chlorine (Cl), a nonmetal, has seven electrons in its outer shell and needs one more to complete the octet. It gains an electron to become Cl^-.

Bond Formation: The oppositely charged ions (cations and anions) are attracted to each other by strong **electrostatic forces**, creating an ionic bond.

Example: Sodium chloride (NaCl) is formed when Na^+ and Cl^- ions come together. The electrostatic attraction between the oppositely charged ions forms a strong bond.

Properties of Ionic Bonds

High melting and boiling points: Ionic compounds typically have high melting and boiling points because the electrostatic forces between ions are very strong, requiring significant energy to overcome.

Brittleness: Ionic compounds tend to be brittle, meaning they break easily when force is applied. This is because when the layers of ions are shifted, like-charged ions are forced together, repelling each other and causing the compound to fracture.

Electrical conductivity: In the solid state, ionic compounds do not conduct electricity because the ions are locked in place. However, when melted or dissolved in water, the ions are free to move, allowing the substance to conduct electricity.

Solubility in water: Many ionic compounds dissolve easily in water because the polar water molecules can interact with and separate the positive and negative ions.

3. Covalent Bond Formation

A **covalent bond** forms when two atoms **share electrons** to achieve stability. This type of bonding is common between **nonmetals** and is a fundamental feature of many organic and biological molecules.

Single, Double, and Triple Bonds: The number of electron pairs shared between two atoms determines the type of covalent bond.

Single bond: One pair of electrons is shared (e.g., H-H in H_2).

Double bond: Two pairs of electrons are shared (e.g., O=O in O_2).

Triple bond: Three pairs of electrons are shared (e.g., N≡N in N_2).

Polar and Nonpolar Covalent Bonds: Covalent bonds can be **polar** or **nonpolar** depending on the **electronegativity** difference between the atoms.

Nonpolar covalent bond: Electrons are shared equally between the atoms because they have similar electronegativities (e.g., H_2 or Cl_2).

Polar covalent bond: Electrons are shared unequally, causing one atom to be slightly negative and the other slightly positive. This occurs when there is a significant difference in electronegativity between the atoms (e.g., H_2O).

Properties of Covalent Bonds

Low melting and boiling points: Covalent compounds often have lower melting and boiling points than ionic compounds because the forces holding the molecules together (intermolecular forces) are weaker than the forces in ionic bonds.

Electrical conductivity: Covalent compounds do not conduct electricity in any state because there are no free ions or electrons to carry a charge.

Solubility: Covalent compounds may or may not dissolve in water, depending on whether they are polar or nonpolar. Polar covalent compounds, like sugar, dissolve in water, whereas nonpolar substances, like oil, do not.

Flexibility: Covalent compounds are often flexible, unlike ionic compounds, which are more rigid and brittle.

4. Metallic Bond Formation

A **metallic bond** is formed when atoms in a metal lattice **pool their electrons**. In this bond, valence electrons are not shared or transferred between specific atoms, but rather they are **delocalized** and move freely throughout the entire structure of the metal.

Electron Pooling: In metallic bonding, atoms of metals contribute their valence electrons to form a "sea of electrons." These electrons are free to move between all the metal atoms, creating a strong bond that holds the metal atoms together.

Example: In copper (Cu), the atoms are held together by metallic bonds. The valence electrons are free to move throughout the copper lattice, giving the metal its distinctive properties.

Properties of Metallic Bonds

Electrical and thermal conductivity: The free movement of electrons in a metal lattice allows metals to conduct both electricity and heat efficiently.

Malleability and ductility: Metals can be hammered into thin sheets (malleable) or drawn into wires (ductile) because the metallic bonds are flexible and allow the atoms to slide past one another without breaking.

Luster: The delocalized electrons in metals interact with light, giving metals their shiny appearance.

High melting and boiling points: Most metals have high melting and boiling points because the metallic bonds are strong, requiring a significant amount of energy to break.

5. Intermolecular Forces

In addition to the three primary types of bonds (ionic, covalent, and metallic), there are **intermolecular forces** that occur between molecules and influence the physical properties of substances.

Hydrogen Bonds: A particularly strong type of dipole-dipole interaction, hydrogen bonding occurs when hydrogen is bonded to highly electronegative atoms like oxygen, nitrogen, or fluorine. This bond is crucial for the properties of water and biological molecules such as DNA.

Example: The hydrogen bonds between water molecules give water its high boiling point and surface tension.

Van der Waals Forces: These are weak forces that arise from temporary shifts in electron density in molecules. Van der Waals forces include:

London Dispersion Forces: Present in all molecules, they arise from temporary dipoles created by momentary shifts in electron positions.

Dipole-Dipole Interactions: Occur between polar molecules, where the positive end of one molecule is attracted to the negative end of another.

6. Bond Strength and Energy

The **strength** of a bond is determined by the amount of energy required to break it. This is referred to as **bond dissociation energy**. Stronger bonds require more energy to break.

Ionic bonds are typically strong because of the electrostatic attraction between oppositely charged ions.

Covalent bonds can vary in strength. Single bonds are generally weaker than double or triple bonds because fewer electrons are shared.

Metallic bonds are strong, but because the electrons are delocalized, metals retain their malleability and ductility even under stress.

Bond **formation** is an **exothermic process**, meaning it releases energy, whereas bond **breaking** is an **endothermic process**, meaning it requires energy.

Understanding bond formation and its properties is essential to the study of chemistry because it explains how atoms combine and the resulting properties of compounds and materials. Whether through the transfer of electrons in ionic bonding, the sharing of electrons in covalent bonding, or the pooling of electrons in metallic bonding, the behavior of atoms is driven by their quest for stability. These bonds influence everything from the hardness of diamonds to the flexibility of metals, and they form the basis for the structure and function of all chemical substances.

Molecular Geometry and VSEPR Theory

Molecular geometry refers to the three-dimensional arrangement of atoms within a molecule. The shape of a molecule is critical in determining its physical properties, reactivity, polarity, and interactions with other molecules. Molecular geometry not only helps chemists understand how molecules interact in chemical reactions but also sheds light on biological functions, material properties, and other molecular behaviors. One of the key tools used to predict and understand molecular shapes is the **VSEPR theory** (Valence Shell Electron Pair Repulsion theory).

1. Importance of Molecular Geometry

The way atoms are arranged in a molecule affects a range of properties:

Reactivity: Some molecular shapes make it easier for molecules to interact or bond with other substances.

Polarity: The distribution of electron density can make a molecule polar or nonpolar, which in turn influences solubility, boiling points, and intermolecular interactions.

Intermolecular forces: The geometry of molecules determines the strength of forces like hydrogen bonding, van der Waals forces, or dipole-dipole interactions.

Biological function: In biological molecules, like proteins or DNA, the specific 3D arrangement is crucial for function. Enzyme activity, for instance, relies on a precise fit between molecules, often described as a "lock and key" mechanism.

2. VSEPR Theory: Understanding Molecular Shapes

The **VSEPR theory** is a model used to predict the geometry of molecules based on the idea that electron pairs around a central atom will arrange themselves as far apart as possible to minimize repulsion. Since electrons are negatively charged and repel each other, the shape of a molecule is largely determined by the repulsion between the bonding and non-bonding electron pairs (lone pairs) around the central atom.

Key Principle: Electron pairs (both bonding pairs and lone pairs) around a central atom will position themselves as far away from each other as possible to minimize repulsive forces.

VSEPR theory is primarily focused on the **electron domains** around the central atom. These electron domains include both bonding pairs (single, double, or triple bonds) and lone pairs. The number of electron domains around a central atom dictates the molecule's basic shape.

3. Steps in Applying VSEPR Theory

To determine the molecular geometry of a molecule using VSEPR theory, follow these steps:

Draw the Lewis Structure: Start by drawing the Lewis structure of the molecule, showing all the bonding and lone pairs of electrons around the central atom.

Count Electron Domains: Count the total number of electron domains around the central atom. An electron domain refers to any bond (single, double, or triple) or lone pair of electrons.

Determine Electron Geometry: The total number of electron domains around the central atom determines the basic shape or electron geometry.

Determine Molecular Geometry: After considering the electron geometry, focus on the positions of atoms (ignoring lone pairs) to find the molecular geometry.

4. Common Electron Geometries and Molecular Shapes

Here are some common electron geometries and molecular geometries derived from VSEPR theory:

Linear Geometry (180° bond angle):

Electron domains: 2

Example: Carbon dioxide (CO_2)

Molecular geometry: If there are no lone pairs, the molecule is linear. CO_2 has two bonding domains and no lone pairs, leading to a straight linear geometry.

Trigonal Planar Geometry (120° bond angle):

Electron domains: 3

Example: Boron trifluoride (BF_3)

Molecular geometry: With three bonding domains and no lone pairs, the atoms form a flat triangle around the central atom.

Bent (V-shaped) Geometry:

Electron domains: 3 (2 bonding pairs, 1 lone pair) or 4 (2 bonding pairs, 2 lone pairs)

Example: Water (H_2O)

Molecular geometry: Lone pairs exert more repulsion than bonding pairs, so the bonding pairs are pushed closer together, creating a bent shape. Water has two bonding pairs and two lone pairs, leading to a bent geometry with a bond angle of approximately 104.5°.

Tetrahedral Geometry (109.5° bond angle):

Electron domains: 4

Example: Methane (CH_4)

Molecular geometry: Four bonding pairs are arranged symmetrically in 3D space to form a tetrahedral shape. CH_4, with no lone pairs, adopts this perfect tetrahedral shape.

Trigonal Pyramidal Geometry:

Electron domains: 4 (3 bonding pairs, 1 lone pair)

Example: Ammonia (NH_3)

Molecular geometry: The lone pair repels the bonding pairs, creating a pyramid-like structure where the three hydrogen atoms form the base and the nitrogen atom is at the apex. NH_3 has a trigonal pyramidal geometry with a bond angle slightly less than 109.5° due to the lone pair repulsion.

Trigonal Bipyramidal Geometry:

Electron domains: 5

Example: Phosphorus pentachloride (PCl_5)

Molecular geometry: In this geometry, three atoms are arranged in a trigonal plane around the central atom, with two additional atoms positioned above and below the plane (axial positions). This arrangement results in bond angles of 90° and 120°.

Octahedral Geometry (90° bond angle):

Electron domains: 6

Example: Sulfur hexafluoride (SF_6)

Molecular geometry: All six electron pairs are bonding pairs, forming an octahedral geometry where the atoms are arranged symmetrically around the central sulfur atom with 90° bond angles.

5. Effects of Lone Pairs on Molecular Geometry

Lone pairs of electrons play a significant role in altering the ideal geometries predicted by VSEPR theory. They occupy more space than bonding pairs because they are localized around the central atom, leading to greater repulsive forces. This results in deviations from ideal bond angles and molecular shapes.

Lone pairs reduce bond angles: For instance, in ammonia (NH_3), the ideal tetrahedral bond angle is 109.5°, but due to the lone pair on nitrogen, the actual bond angle is reduced to around 107°.

Lone pairs create asymmetry: In water (H_2O), the two lone pairs on oxygen push the bonding pairs closer together, creating a bent shape with a bond angle of approximately 104.5° instead of the ideal 109.5°.

6. Molecular Polarity and VSEPR Theory

Molecular geometry has a direct impact on the **polarity** of a molecule. Polarity is determined by the difference in electronegativity between atoms and the distribution of electron density across the molecule.

Symmetrical molecules: If a molecule is symmetrical, even if it contains polar bonds, the bond dipoles cancel each other out, resulting in a **nonpolar molecule**.

Example: Carbon dioxide (CO_2) has polar bonds between carbon and oxygen, but its linear shape ensures the dipoles cancel out, making it nonpolar.

Asymmetrical molecules: In asymmetrical molecules, the dipoles do not cancel, leading to a **polar molecule**.

Example: Water (H_2O) has polar O-H bonds, and the bent geometry prevents the dipoles from canceling out, making water a polar molecule.

7. Limitations of VSEPR Theory

While VSEPR theory is a valuable tool for predicting molecular geometry, it does have some limitations:

It does not account for **electron delocalization**, such as in molecules with resonance structures, where electrons are spread over multiple atoms.

It doesn't fully explain the **bonding in transition metals** or other complex bonding situations, where hybridization and d-orbital interactions become more important.

Quantum mechanical models, like molecular orbital theory, are sometimes required for a more accurate depiction of electron behavior and bonding.

Understanding molecular geometry through VSEPR theory is critical in predicting the shape, polarity, and reactivity of molecules. By arranging electron pairs to minimize repulsion, VSEPR helps chemists visualize the 3D arrangement of atoms within molecules. This, in turn, provides insight into how molecules will interact with each other, their behavior in reactions, and their physical properties. Molecular geometry is not just a theoretical concept; it has practical applications in fields ranging from material science to biochemistry, where the shape of a molecule can dictate its functionality and importance.

Polar and Nonpolar Molecules

The classification of molecules into **polar** and **nonpolar** is fundamental in understanding the behavior of substances in various chemical reactions and physical processes. The distinction between polar and nonpolar molecules directly impacts properties such as solubility, boiling points, melting points, intermolecular forces, and the overall reactivity of the molecule. This differentiation is based on the distribution of electrical charge across the molecule, which is influenced by factors like **electronegativity** and **molecular geometry**.

1. Electronegativity and Bond Polarity

The concept of **electronegativity** is central to understanding polarity. Electronegativity refers to the ability of an atom to attract electrons in a chemical bond. The higher the electronegativity, the more strongly an atom pulls bonding electrons toward itself. Different atoms have different electronegativities, and when two atoms form a bond, the difference in their electronegativity values determines the bond type.

Nonpolar Covalent Bond: When two atoms have identical or very similar electronegativities (such as in molecules made of the same element, like H_2, N_2, or O_2), they share electrons equally. As a result, the bond is considered **nonpolar**, and the electron distribution is even across the molecule.

Polar Covalent Bond: When there is a significant difference in electronegativity between two atoms (for example, in HCl, where chlorine is much more electronegative than hydrogen), the electrons are shared unequally. The more electronegative atom pulls the shared electrons closer, creating a partial negative charge (denoted $\delta-$), while the less electronegative atom gets a partial positive charge (denoted $\delta+$). This unequal sharing leads to a **polar covalent bond**.

In a polar bond, this separation of charges creates a **dipole moment**, a vector quantity that points from the partially positive atom to the partially negative atom. The greater the difference in electronegativity, the stronger the dipole moment, and hence the greater the bond polarity.

2. Polar Molecules

A **polar molecule** is one in which the overall distribution of electrons results in regions of partial positive and partial negative charge. This uneven electron distribution leads to a net dipole moment for the entire molecule. Several factors contribute to whether a molecule is polar:

Bond Polarity: If the individual bonds in the molecule are polar due to differences in electronegativity, the molecule could be polar.

Molecular Geometry: Even if a molecule contains polar bonds, its overall shape can influence whether these dipoles cancel out or reinforce each other. For instance, if the molecular geometry is symmetrical, the polarities of the individual bonds might cancel each other, resulting in a nonpolar molecule. If the geometry is asymmetrical, the dipoles may not cancel, leading to a polar molecule.

Examples of Polar Molecules:

Water (H_2O): Water is a classic example of a polar molecule. The oxygen atom is much more electronegative than hydrogen, creating polar O-H bonds. Moreover, the molecular geometry of water is bent (due to the two lone pairs on oxygen), which prevents the bond dipoles from canceling out. This creates a net dipole moment, making water polar. This polarity is responsible for water's high boiling point, strong hydrogen bonding, and its ability to dissolve many ionic and polar compounds.

Ammonia (NH_3): Ammonia also has a polar nature. The nitrogen atom is more electronegative than hydrogen, leading to polar N-H bonds. Ammonia's trigonal pyramidal shape, with a lone pair on nitrogen, ensures that the dipole moments do not cancel out, resulting in a net dipole and making the molecule polar.

Hydrogen Chloride (HCl): In hydrogen chloride, chlorine is much more electronegative than hydrogen, creating a strong dipole moment in the H-Cl bond. Since the molecule consists of only one bond, the entire molecule is polar.

3. Nonpolar Molecules

In a **nonpolar molecule**, the electron distribution is more uniform, meaning there is no overall dipole moment. Nonpolar molecules can arise from two main scenarios:

Nonpolar Bonds: When the atoms in the molecule have similar or identical electronegativities, the electrons are shared equally, leading to nonpolar bonds. Examples include diatomic molecules like N_2, O_2, and H_2.

Symmetrical Geometry: Even if a molecule contains polar bonds, the overall molecule can still be nonpolar if its geometry is symmetrical. In such cases, the dipoles of individual bonds cancel each other out, resulting in no net dipole moment.

Examples of Nonpolar Molecules:

Methane (CH_4): Methane is a nonpolar molecule despite having polar C-H bonds. The molecule's tetrahedral geometry is symmetrical, causing the dipoles of the C-H bonds to cancel out, making the molecule nonpolar. Methane's nonpolarity explains its low boiling point and poor solubility in water.

Carbon Dioxide (CO_2): Carbon dioxide has two polar C=O bonds. However, the linear geometry of the molecule ensures that the dipoles of the two bonds cancel out, resulting in a nonpolar molecule. This symmetry is why CO_2 is a gas at room temperature and does not dissolve well in polar solvents like water.

Diatomic Molecules (O_2, N_2, H_2): These molecules are made up of two atoms of the same element. Since the electronegativities of the atoms are identical, there is no difference in electron sharing, making these molecules nonpolar.

4. Polarity and Molecular Interactions

The polarity of a molecule has significant implications for how it interacts with other molecules, affecting properties such as:

Intermolecular Forces: Polar molecules interact through dipole-dipole interactions and hydrogen bonding (if hydrogen is bonded to highly electronegative elements like O, N, or F), which are much stronger than the London dispersion forces that govern nonpolar molecules.

Dipole-Dipole Interactions: Polar molecules tend to align so that the positive end of one molecule is attracted to the negative end of another. This leads to stronger intermolecular attractions and higher boiling and melting points.

Hydrogen Bonding: A special type of dipole-dipole interaction occurs in molecules where hydrogen is bonded to highly electronegative atoms like oxygen, nitrogen, or fluorine. This creates a very strong attraction between molecules, as seen in water, leading to higher boiling points and strong cohesive properties.

London Dispersion Forces: In nonpolar molecules, the only significant force between molecules is the London dispersion force, which arises from temporary dipoles created by fluctuations in electron distribution. These forces are generally weaker, leading to lower boiling and melting points for nonpolar substances.

5. Solubility: "Like Dissolves Like"

One of the most significant effects of molecular polarity is its influence on **solubility**. A simple rule of thumb in chemistry is "like dissolves like," meaning polar substances tend to dissolve in polar solvents, while nonpolar substances dissolve in nonpolar solvents.

Polar Solvents: Polar molecules, like water, are excellent solvents for other polar molecules and ionic compounds. This is because the positive and negative ends of polar molecules can interact with the charged ions or polar regions of solutes, breaking them apart and allowing them to dissolve.

Example: Sodium chloride (NaCl) dissolves in water because the polar water molecules surround the positive sodium ions and negative chloride ions, separating them and keeping them in solution.

Nonpolar Solvents: Nonpolar molecules dissolve better in nonpolar solvents like hexane or benzene, where the interaction between solute and solvent is governed by London dispersion forces.

Example: Oil, which is nonpolar, does not mix with water (polar) but dissolves well in nonpolar solvents like hexane.

6. Applications of Polar and Nonpolar Molecules

Biological Systems: In biological molecules like proteins and lipids, the distribution of polar and nonpolar regions influences their function and structure. For example, the polar heads of phospholipids interact with water in cell membranes, while the nonpolar tails avoid water, creating a hydrophobic barrier.

Chemical Reactions: The polarity of reactants can determine how they interact and the types of products formed. Polar reactants are more likely to participate in reactions where polar intermediates are favored, while nonpolar reactants tend to undergo reactions typical of nonpolar environments.

Environmental Impact: The polarity of a molecule can affect how it behaves in the environment. Nonpolar molecules like hydrocarbons are not soluble in water, making them difficult to clean up in oil spills, while polar pollutants may dissolve and spread more easily in aquatic environments.

The distinction between polar and nonpolar molecules is essential for understanding many chemical and physical properties, from solubility to intermolecular forces. By considering electronegativity and molecular geometry, chemists can predict whether a molecule will be polar or nonpolar and thus infer how it will behave in different environments. Polarity plays a crucial role in chemical reactions, biological systems, and industrial applications, making it a foundational concept in chemistry.

Intermolecular Forces

Intermolecular forces are the forces of attraction or repulsion between molecules or ions that occur outside the bonds within molecules. These forces play a crucial role in determining the physical properties of substances, such as boiling points, melting points, viscosity, and solubility. While **intramolecular forces** (such as covalent, ionic, and metallic bonds) hold atoms together within a molecule, intermolecular forces govern the interactions between molecules themselves.

Intermolecular forces are generally weaker than intramolecular bonds but are still significant enough to affect a substance's macroscopic behavior. The strength and type of these forces depend on the structure and polarity of the molecules involved.

1. Types of Intermolecular Forces

Intermolecular forces can be classified into several categories based on their strength and the nature of the interacting particles. The main types are:

London Dispersion Forces (Van der Waals forces)

Dipole-Dipole Interactions

Hydrogen Bonding

Ion-Dipole Forces

Each of these forces plays a specific role in different types of molecules.

2. London Dispersion Forces (Van der Waals Forces)

London dispersion forces (sometimes called Van der Waals forces) are the weakest intermolecular forces. They arise due to temporary fluctuations in the electron distribution within atoms or molecules, which create instantaneous dipoles. These temporary dipoles induce similar dipoles in nearby molecules, leading to weak attractions.

Key Characteristics:

Universality: London dispersion forces occur in **all** molecules, regardless of whether they are polar or nonpolar. However, they are the only intermolecular force present in **nonpolar molecules**.

Dependence on Molecular Size: The strength of dispersion forces increases with the **size and molecular weight** of the molecules. Larger atoms or molecules have more electrons, making it easier for temporary dipoles to form, which enhances the strength of the dispersion forces.

Example: Dispersion forces are responsible for the ability of nonpolar substances like **noble gases** (e.g., helium, neon, argon) and hydrocarbons (e.g., methane, ethane, butane) to condense into liquids at low temperatures. For instance, methane (CH_4), a nonpolar molecule, exhibits only London dispersion forces, and its low boiling point (-161.5°C) reflects the weak nature of these interactions.

Effect on Physical Properties:

Substances with stronger London dispersion forces tend to have **higher boiling and melting points**. This is because it requires more energy to overcome the attraction between larger, heavier molecules with stronger dispersion forces.

For example, **iodine (I_2)** has much stronger dispersion forces than **fluorine (F_2)**, even though both are diatomic molecules. As a result, iodine is a solid at room temperature, while fluorine is a gas.

3. Dipole-Dipole Interactions

Dipole-dipole interactions occur between molecules that have permanent dipoles, meaning that the molecules possess a **net dipole moment** due to differences in electronegativity between atoms in their bonds. In these interactions, the positive end of one polar molecule is attracted to the negative end of another polar molecule.

Key Characteristics:

Polarity: Dipole-dipole interactions only occur between **polar molecules**. The greater the dipole moment, the stronger the dipole-dipole attraction between molecules.

Directional Nature: These interactions are **directional**, meaning the molecules align themselves in such a way that the positive and negative poles face each other, leading to stronger attractions.

Example: **Hydrogen chloride (HCl)** is a polar molecule, with a significant dipole moment because chlorine is much more electronegative than hydrogen. The positive hydrogen end of one HCl molecule is attracted to the negative chlorine end of another, creating dipole-dipole interactions that contribute to HCl's higher boiling point compared to nonpolar molecules of similar size, like methane.

Effect on Physical Properties:

Molecules with significant dipole-dipole interactions tend to have **higher boiling points and melting points** than similar-sized nonpolar molecules because it requires more energy to separate the molecules due to these stronger attractions.

For instance, **acetone (C_3H_6O)**, a polar molecule, has a higher boiling point (56°C) than nonpolar molecules like **propane (C_3H_8)**, even though both have similar molecular weights. The stronger dipole-dipole interactions in acetone account for this difference.

4. Hydrogen Bonding

Hydrogen bonding is a special and highly significant type of dipole-dipole interaction. It occurs when hydrogen is covalently bonded to one of the most **electronegative elements**: **fluorine (F)**, **oxygen (O)**, or **nitrogen (N)**. This results in a highly polarized bond where the hydrogen atom carries a significant partial positive charge ($\delta+$) and is strongly attracted to a lone pair of electrons on a nearby electronegative atom (F, O, or N) in another molecule.

Key Characteristics:

High Strength: Hydrogen bonds are much stronger than regular dipole-dipole interactions, though still weaker than covalent bonds. This is due to the large difference in electronegativity between hydrogen and F, O, or N.

Directional: Like dipole-dipole interactions, hydrogen bonds are highly **directional** and form in a straight line between the hydrogen atom and the electronegative atom in the neighboring molecule.

Example: **Water (H_2O)** is the most well-known example of hydrogen bonding. The oxygen atom in water is highly electronegative and forms strong hydrogen bonds with the hydrogen atoms of neighboring water molecules. These hydrogen bonds are responsible for water's unusually high boiling point (100°C) compared to other molecules of similar size, such as hydrogen sulfide (H_2S), which has no hydrogen bonding and boils at -60°C.

Effect on Physical Properties:

Boiling and Melting Points: Substances that exhibit hydrogen bonding typically have **much higher boiling and melting points** than those that do not. Water's high boiling point and its solid form (ice) being less dense than its liquid form are both due to the strong hydrogen bonds between water molecules.

Viscosity and Surface Tension: Hydrogen bonding also contributes to the **high viscosity** and **surface tension** of substances. For example, water has a high surface tension due to the strong hydrogen bonds at the surface, which create a "film" that allows small objects to float.

Biological Importance: Hydrogen bonding is crucial in biological systems. It stabilizes the structures of **proteins** and **DNA**. In DNA, hydrogen bonds between nitrogenous bases hold the two strands of the double helix together.

5. Ion-Dipole Forces

Ion-dipole forces are the strongest type of intermolecular forces. These occur when an **ion** (either a cation or an anion) interacts with the **dipole** of a polar molecule. Ion-dipole interactions are particularly important in the dissolution of ionic compounds in polar solvents like water.

Key Characteristics:

Strength: Ion-dipole forces are significantly stronger than dipole-dipole interactions because the attraction between a full charge (ion) and a partial charge (dipole) is much stronger than the attraction between two partial charges.

Example: When **sodium chloride (NaCl)** dissolves in water, the positive sodium ions (Na^+) are attracted to the negative oxygen ends of the water molecules, while the negative chloride ions (Cl^-) are attracted to the positive hydrogen ends. These ion-dipole forces are responsible for pulling the ions away from the crystal lattice and allowing them to dissolve in the solvent.

Effect on Physical Properties:

Solubility: Ion-dipole interactions are crucial for the solubility of ionic compounds in polar solvents. The ability of water to dissolve salts, acids, and bases is a direct result of these strong interactions.

Electrolyte Solutions: Solutions containing ions, such as saline water or electrolyte solutions, owe their conductivity and many other properties to ion-dipole interactions.

6. Comparison of Intermolecular Forces

The strength of intermolecular forces can be ranked as follows, from weakest to strongest:

London Dispersion Forces: Weakest, present in all molecules but dominant in nonpolar molecules.

Dipole-Dipole Interactions: Moderate strength, only in polar molecules.

Hydrogen Bonding: Stronger than dipole-dipole, specific to molecules where hydrogen is bonded to F, O, or N.

Ion-Dipole Forces: Strongest, occurring between ions and polar molecules, especially in solutions.

7. Role of Intermolecular Forces in Everyday Life

Boiling and Melting Points: Stronger intermolecular forces result in higher boiling and melting points. Substances with hydrogen bonds, like water, require more energy (heat) to overcome these interactions.

Viscosity: Substances with stronger intermolecular forces tend to be more viscous. For example, glycerol, which has multiple hydrogen bonds, is much more viscous than water.

Solubility: The "like dissolves like" rule is largely based on intermolecular forces. Polar substances dissolve in polar solvents (due to dipole-dipole or hydrogen bonding), while nonpolar substances dissolve in nonpolar solvents (due to London dispersion forces).

Biological Functionality: Hydrogen bonding is essential in biological systems, particularly in the structure and function of DNA and proteins.

Intermolecular forces govern the behavior of molecules beyond their individual bonds, playing a critical role in determining the physical properties of substances. By understanding the different types of intermolecular forces—London dispersion forces, dipole-dipole interactions, hydrogen bonding, and ion-dipole forces—chemists can predict how substances will behave in various contexts, from boiling to dissolving, and from biological functions to industrial applications. These forces, though weaker than covalent or ionic bonds, are key to the interactions that shape our world.

Chapter 6: Chemical Reactions

Understanding Chemical Reactions

Chemical reactions are at the heart of chemistry and represent the process by which substances interact to form new substances. During a chemical reaction, the bonds between atoms in the reactants are broken, rearranged, and new bonds are formed, resulting in the creation of products with different properties from the original substances.

1. What is a Chemical Reaction?

A **chemical reaction** is a process that involves the **transformation of one or more substances** into new substances with distinct physical and chemical properties. These transformations occur at the molecular level, where the atoms and molecules of the reactants undergo rearrangement.

The substances present at the beginning of a chemical reaction are called **reactants**, and the substances formed as a result of the reaction are called **products**.

For example, when hydrogen (H_2) reacts with oxygen (O_2) to form water (H_2O), the process can be represented as:

$$2H_2 + O_2 \rightarrow 2H_2O$$

In this reaction:

Hydrogen and oxygen are the **reactants**.

Water is the **product**.

2. Signs of a Chemical Reaction

Several observable signs can indicate that a chemical reaction has occurred. While these signs do not always guarantee that a reaction is chemical (as some can also accompany physical changes), they provide strong clues that a new substance has been formed. Common signs include:

Change in color: A noticeable color change suggests that new substances with different colors are being formed. For example, the rusting of iron is marked by a color change from metallic grey to reddish-brown.

Formation of a gas: If bubbles or fumes are produced, this often indicates a chemical reaction. For instance, mixing baking soda with vinegar produces carbon dioxide gas.

Formation of a precipitate: When a solid forms in a liquid solution during a reaction, it is called a precipitate. An example is when silver nitrate and sodium chloride solutions mix, forming a white solid of silver chloride.

Temperature change: Many chemical reactions involve the release or absorption of energy, often in the form of heat. Exothermic reactions release heat (e.g., combustion), while endothermic reactions absorb heat (e.g., the dissolving of ammonium nitrate in water).

Change in odor: A new smell often indicates the formation of different chemicals. This is common in reactions involving organic substances, such as the sour odor from the fermentation of sugar into alcohol and vinegar.

3. Types of Chemical Reactions

Chemical reactions are broadly categorized based on the nature of the reactants and the products, as well as how bonds are broken and formed. Some of the main types of reactions include:

A. Synthesis Reactions (Combination Reactions)

In a **synthesis reaction**, two or more reactants combine to form a single product. This is one of the simplest types of chemical reactions.

General form: $A + B \rightarrow AB$

Example: When sodium (Na) reacts with chlorine (Cl_2), it forms sodium chloride (NaCl):

$$2Na + Cl_2 \rightarrow 2NaCl$$

B. Decomposition Reactions

In a **decomposition reaction**, a single compound breaks down into two or more simpler substances. Decomposition reactions are often initiated by heat, light, or electricity.

General form: $AB \rightarrow A + B$

Example: When calcium carbonate ($CaCO_3$) is heated, it decomposes into calcium oxide (CaO) and carbon dioxide (CO_2): $CaCO_3 \rightarrow CaO + CO_2$

C. Single Displacement Reactions (Single Replacement Reactions)

In a **single displacement reaction**, an element replaces another element in a compound, leading to the formation of a new element and a new compound.

General form: $A + BC \rightarrow AC + B$

Example: When zinc (Zn) reacts with hydrochloric acid (HCl), it displaces hydrogen, forming zinc chloride ($ZnCl_2$) and hydrogen gas (H_2): $Zn + 2HCl \rightarrow ZnCl_2 + H_2$

D. Double Displacement Reactions (Double Replacement Reactions)

In a **double displacement reaction**, the cations and anions of two different compounds swap places, forming two new compounds. These reactions often occur in aqueous solutions and may result in the formation of a precipitate, gas, or neutral compound (such as water).

General form: $AB + CD \rightarrow AD + CB$

Example: When silver nitrate ($AgNO_3$) reacts with sodium chloride (NaCl), silver chloride (AgCl) precipitates, and sodium nitrate ($NaNO_3$) remains in solution:

$$AgNO_3 + NaCl \rightarrow AgCl + NaNO_3$$

E. Combustion Reactions

In a **combustion reaction**, a substance reacts with oxygen, usually releasing energy in the form of heat and light. Combustion is typically associated with the burning of hydrocarbons (compounds containing hydrogen and carbon), which produce carbon dioxide and water as products.

General form: $Hydrocarbon + O_2 \rightarrow CO_2 + H_2O$

Example: The combustion of methane (CH_4) in oxygen produces carbon dioxide and water:

$$CH_4 + 2O_2 \rightarrow CO_2 + 2H_2O$$

4. Energy in Chemical Reactions: Exothermic vs. Endothermic

Chemical reactions involve the breaking and forming of bonds, which requires energy. Depending on the energy dynamics, reactions are classified as either **exothermic** or **endothermic**.

A. Exothermic Reactions

In an **exothermic reaction**, energy is released into the surroundings, usually in the form of heat, light, or sound. This makes the surroundings warmer, and the energy content of the products is lower than that of the reactants.

Example: The combustion of fuels, such as the burning of wood or gasoline, is an exothermic process:

$$C_6H_{12}O_6 + 6O_2 \rightarrow 6CO_2 + 6H_2O + \text{energy (heat)}$$

B. Endothermic Reactions

In an **endothermic reaction**, energy is absorbed from the surroundings, which lowers the temperature of the environment. The products of an endothermic reaction have higher energy content than the reactants.

Example: Photosynthesis is an endothermic process in which plants absorb sunlight to convert carbon dioxide and water into glucose and oxygen:

$$6CO_2 + 6H_2O + \text{energy (sunlight)} \rightarrow C_6H_{12}O_6 + 6O_2$$

5. The Law of Conservation of Mass

One of the fundamental principles of chemical reactions is the **Law of Conservation of Mass**, which states that mass is neither created nor destroyed in a chemical reaction. The total mass of the reactants must equal the total mass of the products. This principle is essential when balancing chemical equations to ensure that the number of atoms for each element is the same on both sides of the equation.

Example: In the synthesis of water, two molecules of hydrogen gas (H_2) react with one molecule of oxygen gas (O_2) to produce two molecules of water (H_2O). The total number of hydrogen and oxygen atoms remains the same before and after the reaction:

$$2H_2 + O_2 \rightarrow 2H_2O$$

6. Catalysts and Reaction Rates

In many chemical reactions, the rate at which the reaction occurs can be influenced by various factors, such as temperature, pressure, and concentration of reactants. **Catalysts** are substances that speed up a chemical reaction without being consumed in the process. Catalysts work by lowering the activation energy required for the reaction, allowing it to proceed more quickly.

Example: Enzymes are biological catalysts that accelerate reactions in living organisms. For instance, the enzyme **amylase** helps break down starches into sugars during digestion.

7. Importance of Understanding Chemical Reactions

Understanding chemical reactions is fundamental to nearly every aspect of chemistry. From industrial processes like the production of fertilizers and pharmaceuticals to everyday activities like cooking and digestion, chemical reactions are constantly occurring around us. Studying how and why these reactions take place allows chemists to predict outcomes, design new materials, develop energy sources, and improve environmental sustainability.

In summary, chemical reactions are the processes that lead to the transformation of substances through the breaking and forming of chemical bonds. They are governed by principles such as the conservation of mass and energy changes and are classified into different types based on the nature of the reactants and products. Understanding these processes is critical for advancing both scientific knowledge and practical applications.

Types of Chemical Reactions

Chemical reactions, the processes by which substances transform into different substances, are the foundation of all chemical processes. While there are countless chemical reactions occurring in nature and in the laboratory, they can be broadly categorized into several distinct types based on how the reactants interact and how products are formed. Understanding these different types of chemical reactions is essential for predicting the behavior of chemicals and their products.

Here are the main types of chemical reactions:

1. Synthesis Reactions (Combination Reactions)

A **synthesis reaction** occurs when two or more simple substances combine to form a more complex product. In these reactions, atoms or molecules bond together to create a single compound. Synthesis reactions are fundamental in chemistry and are often seen in the formation of compounds from their elements.

General form: $A + B \rightarrow AB$

Example: The combination of sodium (Na) and chlorine (Cl_2) to form sodium chloride (NaCl):

$$2Na + Cl_2 \rightarrow 2NaCl$$

Synthesis reactions are common in both industrial chemistry and biological processes. For instance, in the body, the synthesis of proteins from amino acids is a vital reaction for building muscle and tissue.

2. Decomposition Reactions

In a **decomposition reaction**, a single compound breaks down into two or more simpler substances. This is essentially the reverse of a synthesis reaction. Decomposition reactions usually require an external source of energy, such as heat, light, or electricity, to break the bonds within the compound.

General form: $AB \rightarrow A + B$

Example: The breakdown of water (H_2O) into hydrogen gas (H_2) and oxygen gas (O_2) when electricity is passed through it: $2H_2O \rightarrow 2H_2 + O_2$

Decomposition reactions are crucial in processes like the breakdown of organic matter during decay and in industrial applications such as the extraction of metals from their ores.

3. Single Displacement Reactions (Single Replacement Reactions)

In a **single displacement reaction**, one element replaces another element in a compound. This occurs when a more reactive element displaces a less reactive element from a compound, forming a new compound and releasing the displaced element.

General form: $A + BC \rightarrow AC + B$

Example: Zinc (Zn) displaces hydrogen (H) in hydrochloric acid (HCl) to form zinc chloride ($ZnCl_2$) and hydrogen gas (H_2): $Zn + 2HCl \rightarrow ZnCl_2 + H_2$

Single displacement reactions are common in metal reactivity series, where more reactive metals displace less reactive metals from their compounds. This type of reaction is also seen in redox processes, where electrons are transferred between elements.

4. Double Displacement Reactions (Double Replacement Reactions)

A **double displacement reaction** occurs when the cations (positively charged ions) and anions (negatively charged ions) of two different compounds exchange places, forming two new compounds. These reactions often take place in aqueous solutions, where one of the products might be a precipitate, a gas, or water.

General form: $AB + CD \rightarrow AD + CB$

Example: When silver nitrate ($AgNO_3$) reacts with sodium chloride (NaCl), silver chloride (AgCl) precipitates, and sodium nitrate ($NaNO_3$) remains in solution: $AgNO_3 + NaCl \rightarrow AgCl + NaNO_3$

Double displacement reactions are commonly used in qualitative analysis in chemistry labs to detect the presence of particular ions in a solution. They also play a significant role in biological systems, such as in the neutralization of acids and bases.

5. Combustion Reactions

A **combustion reaction** is a type of exothermic reaction that occurs when a substance reacts rapidly with oxygen, releasing energy in the form of heat and light. Combustion typically involves organic compounds, particularly hydrocarbons, and always results in the formation of carbon dioxide (CO_2) and water (H_2O) as products.

General form: $Fuel + O_2 \rightarrow CO_2 + H_2O + \text{energy (heat)}$

Example: The combustion of methane (CH_4) in oxygen produces carbon dioxide, water, and energy:

$$CH_4 + 2O_2 \rightarrow CO_2 + 2H_2O + \text{energy}$$

Combustion reactions are crucial for the generation of energy, whether in the burning of fossil fuels for electricity, the internal combustion engines of vehicles, or even the metabolic processes in living organisms.

6. Redox Reactions (Oxidation-Reduction Reactions)

Redox reactions are reactions in which electrons are transferred from one substance to another. These reactions involve two key processes: oxidation (the loss of electrons) and reduction (the gain of electrons). Both processes occur simultaneously, as one substance is oxidized while another is reduced.

General form: Oxidation: $A \rightarrow A^{n+} + ne^-$ Reduction: $B^{n+} + ne^- \rightarrow B$

Example: In the reaction between sodium (Na) and chlorine (Cl_2), sodium is oxidized to form Na^+ ions, and chlorine is reduced to form Cl^- ions: $2Na + Cl_2 \rightarrow 2NaCl$

Redox reactions are essential in energy production, especially in electrochemical cells like batteries and fuel cells. They also play a key role in biological processes like respiration and photosynthesis.

7. Neutralization Reactions

A **neutralization reaction** is a specific type of double displacement reaction where an acid reacts with a base to produce a salt and water. This reaction involves the transfer of a proton (H^+) from the acid to the base.

General form: $Acid + Base \rightarrow Salt + Water$

Example: When hydrochloric acid (HCl) reacts with sodium hydroxide (NaOH), sodium chloride (NaCl) and water (H_2O) are formed: $HCl + NaOH \rightarrow NaCl + H_2O$

Neutralization reactions are critical in many real-world applications, including the treatment of acid reflux in the human stomach (using antacids) and the neutralization of acidic or basic industrial waste.

8. Precipitation Reactions

A **precipitation reaction** is a type of reaction in which two aqueous solutions react to form an insoluble solid called a precipitate. This occurs because the product of the reaction has low solubility in water.

General form: $AB(aq) + CD(aq) \rightarrow AD(s) + CB(aq)$

Example: When a solution of barium chloride ($BaCl_2$) is mixed with a solution of sulfuric acid (H_2SO_4), barium sulfate ($BaSO_4$) precipitates out of solution as a white solid:

$$BaCl_2(aq) + H_2SO_4(aq) \rightarrow BaSO_4(s) + 2HCl(aq)$$

Precipitation reactions are essential for purifying compounds and removing unwanted ions from solutions. They are used extensively in water treatment and analytical chemistry.

9. Acid-Base Reactions

In **acid-base reactions**, an acid donates a proton (H^+) to a base. These reactions are a fundamental type of chemical interaction and are responsible for maintaining pH balance in biological systems, industrial processes, and natural environments.

General form: $HA + B \rightarrow A^- + BH^+$

Example: The reaction between acetic acid (CH_3COOH) and ammonia (NH_3) forms ammonium acetate:

$$CH_3COOH + NH_3 \rightarrow CH_3COONH_4$$

Acid-base reactions are involved in everything from digestive processes in humans to the regulation of chemical conditions in soil for agriculture.

Understanding the different types of chemical reactions—synthesis, decomposition, single displacement, double displacement, combustion, redox, neutralization, precipitation, and acid-base reactions—provides a foundation for mastering chemical interactions. Each type of reaction involves specific processes, reactants, and products, enabling chemists to predict outcomes and design new reactions for practical applications in science, industry, and everyday life. From energy production to the synthesis of new materials, the study of chemical reactions plays a crucial role in shaping the modern world.

Balancing chemical equations is a critical skill in chemistry, as it ensures that chemical reactions comply with the law of conservation of mass. This law states that matter cannot be created or destroyed in a chemical reaction. Therefore, the number of atoms of each element must remain the same on both sides of a chemical equation.

A balanced chemical equation shows the relationship between reactants (the substances that undergo change) and products (the substances that result from the reaction) in correct proportions. Balancing these equations ensures that the mass and charge are conserved throughout the reaction, reflecting how atoms are rearranged rather than created or destroyed.

Why Balancing Equations Is Important

Balancing equations is essential for several reasons:

Law of Conservation of Mass: This law states that in a closed system, mass must remain constant over time. Therefore, the total mass of reactants must equal the total mass of the products.

Accurate Quantitative Predictions: To understand how much of each substance is needed or will be produced in a reaction, the equation must be balanced. This allows chemists to calculate the proper amounts (stoichiometry) of reactants and products.

Reaction Feasibility: A balanced equation indicates whether a chemical reaction can occur. Without balancing, it would be impossible to tell if a reaction makes sense from a scientific perspective.

Chemical Analysis: Understanding chemical equations helps predict outcomes in both laboratory and industrial processes, ensuring safety and efficiency.

Steps to Balancing a Chemical Equation

Balancing chemical equations can seem challenging at first, but following a systematic approach simplifies the process. Below are the key steps:

1. Write the Unbalanced Equation

The first step in balancing an equation is to write down the unbalanced version. This means listing the reactants and products, often based on a word equation.

For example:

$$\text{Methane} + \text{Oxygen} \rightarrow \text{Carbon dioxide} + \text{Water}$$

In symbolic form, this equation becomes:

$$CH_4 + O_2 \rightarrow CO_2 + H_2O$$

At this stage, the equation is unbalanced because the number of atoms of each element on the left side (reactants) does not match the number on the right side (products).

2. Identify Each Element

List the number of atoms for each element present in both the reactants and products. Using the above example:

On the left (reactants):

$$\mathrm{C} = 1, \mathrm{H} = 4, \mathrm{O} = 2$$

On the right (products):

$$\mathrm{C} = 1, \mathrm{H} = 2, \mathrm{O} = 3$$

(since there are 2 oxygens in CO_2 and 1 in H_2O)

The equation is clearly unbalanced because there are unequal numbers of hydrogen and oxygen atoms on both sides.

3. Balance One Element at a Time

The most straightforward method for balancing chemical equations is to start with the element that appears in only one reactant and one product, which makes it easier to adjust its quantity.

In this case, carbon (C) is already balanced (1 atom on both sides).

Next, balance hydrogen (H). There are 4 hydrogen atoms in the reactants (in CH_4) and 2 hydrogen atoms in the products (in H_2O). To balance hydrogen, multiply H2OH_2O H_2O by 2:

$$CH_4 + O_2 \rightarrow CO_2 + 2H_2O$$

Now, oxygen is also balanced with 4 atoms on both sides.

4. Double-Check the Balancing

Finally, go back and check that the number of atoms for each element is equal on both sides of the equation:

Carbon (C): 1 atom on both sides.

Hydrogen (H): 4 atoms on both sides.

Oxygen (O): 4 atoms on both sides.

The equation is now balanced:

$$CH_4 + 2O_2 \rightarrow CO_2 + 2H_2O$$

5. Use Coefficients, Not Subscripts

It's crucial to understand that when balancing chemical equations, you should only change the coefficients (the numbers in front of molecules), not the subscripts (the numbers within the formulas). Changing subscripts would alter the actual chemical substances, which is not allowed. For instance, altering H_2O to H_2O_2 would create a different chemical (hydrogen peroxide).

Tips for Balancing Complex Equations

Some reactions involve more complicated molecules or multiple reactants and products, making balancing more challenging. Here are some tips for handling such cases:

Balance Polyatomic Ions as a Unit: If the same polyatomic ion appears on both sides of the equation, treat it as a single unit to simplify balancing.

Example:

$$Ba(NO_3)_2 + H_2SO_4 \rightarrow BaSO_4 + HNO_3$$

Balance the nitrate ions (NO_3^-) as a single unit instead of separately balancing nitrogen and oxygen.

Balance Hydrogen and Oxygen Last: In many reactions, hydrogen and oxygen appear in multiple compounds. It's often easiest to balance these elements after all others have been balanced.

Check Charges in Redox Reactions: When balancing redox (oxidation-reduction) reactions, make sure to balance both the atoms and the charges. This often requires adding electrons or using the half-reaction method.

Trial and Error: Some equations are best solved by trial and error, adjusting coefficients as you go. Don't be afraid to try different combinations until the equation is balanced.

Example of a More Complex Equation: Combustion of Ethanol

Let's consider the combustion of ethanol (C_2H_5OH) in oxygen:

$$C_2H_5OH + O_2 \rightarrow CO_2 + H_2O$$

Balance carbon: There are 2 carbon atoms in ethanol, so multiply CO_2 by 2:

$$C_2H_5OH + O_2 \rightarrow 2CO_2 + H_2O$$

Balance hydrogen: There are 6 hydrogen atoms in ethanol, so multiply H_2O by 3:

$$C_2H_5OH + O_2 \rightarrow 2CO_2 + 3H_2O$$

Balance oxygen: There are 7 oxygen atoms on the product side (4 in $2CO_2$ and 3 in $3H_2O$), so adjust O_2 accordingly: $C_2H_5OH + 3O_2 \rightarrow 2CO_2 + 3H_2O$

The equation is now balanced.

Balancing chemical equations is a fundamental aspect of chemistry that ensures the principles of the law of conservation of mass are upheld. By adjusting the coefficients, chemists ensure that the number of atoms of each element remains the same on both sides of the reaction. Following a systematic approach—starting with the unbalanced equation, identifying the number of atoms, and balancing each element—makes the process manageable, even for complex reactions. Mastering this skill is essential for solving stoichiometric problems, predicting the outcome of reactions, and understanding the quantitative relationships in chemical processes.

The Role of Catalysts and Enzymes

In the realm of chemical reactions, the rate at which reactions occur can vary greatly. Some reactions proceed rapidly, while others may take hours, days, or even years. To manipulate and control these reaction rates, chemists often rely on substances known as **catalysts**. Catalysts play a vital role in both laboratory experiments and industrial processes by accelerating reaction rates without being consumed or permanently altered during the reaction.

Among catalysts, **enzymes** are specialized biological molecules that serve as nature's catalysts. They are essential for life, facilitating nearly every biochemical reaction in organisms. Understanding the roles of both catalysts and enzymes is crucial for grasping how reactions can be controlled and optimized.

What Is a Catalyst?

A **catalyst** is a substance that increases the rate of a chemical reaction without being consumed or permanently changed by the reaction. It achieves this by providing an alternative reaction pathway with a lower **activation energy**—the minimum energy required for a reaction to take place. By lowering this energy barrier, catalysts allow more reactant molecules to collide with enough energy to react, which speeds up the process.

Key Characteristics of Catalysts:

Unchanged After the Reaction: A catalyst is not consumed in the chemical reaction. It may participate in intermediate stages, but it will regenerate by the end of the process, ready to catalyze another reaction.

Does Not Alter Equilibrium: While a catalyst accelerates the rate at which equilibrium is reached, it does not affect the position of the equilibrium itself. The amounts of reactants and products at equilibrium remain the same.

Lowers Activation Energy: Catalysts lower the energy required to start a reaction, thereby increasing the number of effective collisions between reactant molecules.

Reaction-Specific: Catalysts are typically selective and may only work for specific reactions. This selectivity allows for precise control over industrial and biological processes.

How Catalysts Work

Catalysts work by providing an **alternative pathway** for the reaction, which has a lower activation energy than the uncatalyzed pathway. Activation energy is the energy barrier that reactants must overcome to form products. When a catalyst is present, the energy required to reach the transition state (the high-energy, unstable state between reactants and products) is lowered. This results in more molecules having enough energy to react at a given temperature, thereby increasing the reaction rate.

Mechanism of Action:

Adsorption: In many cases, particularly with solid catalysts, the reactants are adsorbed onto the surface of the catalyst. Adsorption brings the reactants into close proximity, making it easier for them to collide and react.

Formation of Intermediates: Catalysts often form temporary intermediate compounds with the reactants, lowering the activation energy. Once the products form, the intermediates break down, regenerating the catalyst.

Desorption: After the reaction is complete, the products detach (desorb) from the catalyst, leaving it ready to catalyze further reactions.

Types of Catalysts

Catalysts are categorized into two main types:

1. Homogeneous Catalysts:

These catalysts are in the same phase (solid, liquid, or gas) as the reactants. A common example of homogeneous catalysis is the use of sulfuric acid to catalyze the esterification of carboxylic acids and alcohols.

Example:

Esterification:

$$\mathrm{RCOOH} + \mathrm{ROH} \rightarrow \mathrm{RCOOR} + H_2O$$

In this reaction, sulfuric acid H_2SO_4 acts as a catalyst, helping the reaction proceed at a faster rate in the same liquid phase.

2. Heterogeneous Catalysts:

In this case, the catalyst is in a different phase from the reactants. Solid catalysts in contact with gaseous or liquid reactants are common in industrial processes, such as in the **Haber process** for ammonia production and in catalytic converters in automobiles.

Example:

Haber Process (synthesis of ammonia):

$$N_2(g) + 3H_2(g) \xrightarrow{\text{Fe (s)}} 2NH_3(g)$$

Here, iron (Fe) acts as a solid catalyst, facilitating the reaction between nitrogen and hydrogen gases.

Applications of Catalysts in Industry

Catalysts are indispensable in industrial chemistry, where they are used to enhance reaction rates and improve efficiency. The use of catalysts reduces the energy consumption of processes, leading to cost savings and environmental benefits. Some key industrial applications include:

Haber Process: This process synthesizes ammonia from nitrogen and hydrogen using an iron catalyst, which is critical for producing fertilizers.

Catalytic Converters: Automobiles use catalytic converters containing platinum, palladium, and rhodium to convert harmful gases such as carbon monoxide and nitrogen oxides into less harmful substances like carbon dioxide and nitrogen.

Petroleum Refining: Catalysts play a role in **cracking** heavy hydrocarbon molecules into smaller, more useful ones like gasoline and diesel, which are vital for transportation.

Enzymes: Biological Catalysts

Enzymes are biological macromolecules, usually proteins, that function as highly efficient and specific catalysts in biochemical reactions. They are essential for all living organisms, speeding up reactions that would otherwise occur too slowly to sustain life. Enzymes operate under mild conditions, such as body temperature and physiological pH, making them remarkably efficient.

Key Characteristics of Enzymes:

Highly Specific: Each enzyme catalyzes a specific reaction or set of closely related reactions. This specificity arises from the unique 3D structure of the enzyme, particularly the **active site**, where the substrate binds.

Speed: Enzymes can accelerate reactions by factors of up to a billion times or more. This efficiency is crucial for metabolic processes that occur within living cells.

Mild Conditions: Enzymes work under the mild conditions of temperature and pH that are typical of biological systems, unlike many chemical catalysts that require high temperatures and pressures.

Regulated Activity: Enzyme activity can be regulated by the cell through various mechanisms, such as **feedback inhibition**, allowing cells to control metabolic pathways dynamically.

How Enzymes Work

Enzymes work by binding to a specific reactant or **substrate** at the enzyme's **active site**. The active site has a specific shape and chemical environment that promotes the transformation of the substrate into the product. The enzyme itself is not consumed in the reaction and can catalyze the transformation of many substrate molecules over time.

Substrate Binding: The substrate fits into the enzyme's active site, like a key fitting into a lock (this is known as the "lock-and-key" model, although a more accurate model is the "induced fit," where the enzyme slightly changes shape to accommodate the substrate).

Catalysis: The enzyme lowers the activation energy for the reaction, often by stabilizing the transition state or bringing reactants into closer proximity.

Product Formation: The reaction proceeds, and the product is formed.

Enzyme Release: The product is released from the active site, leaving the enzyme ready to bind more substrate.

Enzyme Example: Amylase

Amylase is an enzyme that breaks down starch into simpler sugars during digestion. It is found in saliva and the pancreas, helping to accelerate the breakdown of carbohydrates into glucose, which the body can absorb and use for energy.

Factors Affecting Enzyme Activity

Several factors can affect the rate at which enzymes catalyze reactions:

Temperature: Enzymes have an optimal temperature range. Beyond this range, the enzyme may denature (lose its structure), and its activity will decline.

pH: Each enzyme has an optimal pH range. Extreme pH levels can disrupt the ionic bonds in the enzyme, altering its shape and reducing its activity.

Substrate Concentration: As substrate concentration increases, the rate of reaction increases until the enzyme becomes saturated. At saturation, all enzyme active sites are occupied, and adding more substrate won't increase the reaction rate.

Inhibitors: Enzyme activity can be reduced by inhibitors, which are molecules that bind to the enzyme and decrease its activity. Inhibitors can be **competitive** (competing with the substrate for the active site) or **non-competitive** (binding elsewhere on the enzyme, changing its shape and functionality).

Industrial and Medical Applications of Enzymes

Enzymes are also used in various industrial processes and medical treatments:

Detergents: Enzymes like proteases and lipases are added to laundry detergents to break down proteins and fats in stains.

Food Industry: Enzymes are used to produce products like cheese, beer, and high-fructose corn syrup by accelerating fermentation and other processes.

Medical Diagnostics: Enzymes are used in tests to diagnose conditions like diabetes (e.g., glucose oxidase in blood sugar testing).

Biopharmaceuticals: Enzymes are used to manufacture certain drugs and in therapies such as enzyme replacement therapy for genetic disorders like Gaucher's disease.

Catalysts and enzymes are indispensable in accelerating chemical reactions, both in industrial processes and biological systems. Catalysts lower the activation energy required for reactions to proceed, increasing the reaction rate without being consumed in the process. Enzymes, as biological catalysts, are highly specific and efficient, playing a crucial role in regulating biochemical reactions essential for life. Whether in the lab, industry, or within living cells, understanding the function and mechanisms of catalysts and enzymes is critical for controlling and optimizing chemical reactions.

Stoichiometry and Limiting Reactants

Stoichiometry and **limiting reactants** are critical concepts in understanding chemical reactions. They help chemists predict the amounts of products formed and the efficiency of reactions. By mastering these concepts, students and chemists can calculate yields, optimize resource use, and analyze reactions more effectively.

Stoichiometry: The Arithmetic of Chemistry

Stoichiometry is the calculation of the quantities of reactants and products in chemical reactions. The word comes from the Greek "stoicheion," meaning element, and "metron," meaning measure. It allows chemists to determine how much of each substance is involved in a reaction based on a balanced chemical equation.

In every chemical reaction, atoms are neither created nor destroyed; they are merely rearranged. A balanced chemical equation shows the proportion of molecules or moles of each reactant and product involved, which is the foundation of stoichiometric calculations.

Key Concepts in Stoichiometry

Mole Ratios: The coefficients in a balanced chemical equation represent the ratios in which substances react and are produced. These ratios, called **mole ratios**, are essential for stoichiometric calculations. For example, in the reaction:

$$2H_2(g) + O_2(g) \rightarrow 2H_2O(l)$$

The coefficients (2 for H_2, 1 for O_2, and 2 for H_2O) tell us that 2 moles of hydrogen gas react with 1 mole of oxygen gas to produce 2 moles of water. This relationship is used to calculate the amounts of reactants and products in any similar reaction.

Molar Mass: The molar mass of a substance (in grams per mole) is essential in stoichiometry, as it allows the conversion between the mass of a substance and the number of moles. For example, the molar mass of water

H_2O • is 18.02 g/mol, meaning one mole of water weighs 18.02 grams.

Balanced Chemical Equations: A balanced equation ensures the law of conservation of mass is obeyed, meaning the number of atoms of each element is the same on both sides of the equation. Without balancing, stoichiometric calculations will give incorrect results.

The Mole Concept: Stoichiometry relies heavily on the concept of the mole, which is a counting unit for atoms, molecules, ions, or other entities. One mole contains 6.022×10^{23} entities (Avogadro's number). Using the mole as a basis, stoichiometry allows for the scaling of reactions from the microscopic (atomic) to the macroscopic (gram) scale.

Steps for Performing Stoichiometric Calculations

Write the balanced chemical equation for the reaction.

Determine the molar masses of the reactants and products.

Convert quantities of reactants or products (usually given in grams) to moles using their molar mass.

Use mole ratios from the balanced equation to relate the number of moles of one substance to the number of moles of another.

Convert moles back to grams if necessary, using molar masses.

Example: Stoichiometric Calculation

Let's consider the combustion of methane:

$$CH_4(g) + 2O_2(g) \rightarrow CO_2(g) + 2H_2O(l)$$

If you have 16.0 g of methane CH_4, how many grams of oxygen O_2 are required for complete combustion?

Step 1: The balanced equation shows that 1 mole of CH_4 reacts with 2 moles of O_2.

Step 2: Find the molar masses: CH_4 = 16.04 g/mol, O_2 = 32.00 g/mol.

Step 3: Convert the mass of CH_4 to moles:

$$0.9975 \, \text{mol} \, CH_4 \times \frac{2 \, \text{mol} \, O_2}{1 \, \text{mol} \, CH_4} = 1.995 \, \text{mol} \, O_2$$

Step 4: Use the mole ratio to find moles of O_2:

$$0.9975 \, \text{mol} \, CH_4 \times \frac{2 \, \text{mol} \, O_2}{1 \, \text{mol} \, CH_4} = 1.995 \, \text{mol} \, O_2$$

Step 5: Convert moles of O_2 to grams:

$$1.995 \, \text{mol} \, O_2 \times 32.00 \, \text{g/mol} = 63.84 \, \text{g} \, O_2$$

So, you need 63.84 g of oxygen to completely combust 16.0 g of methane.

Limiting Reactants: Controlling the Reaction Outcome

In a chemical reaction, the **limiting reactant** (or limiting reagent) is the substance that is completely consumed first, limiting the amount of product that can be formed. Once the limiting reactant is used up, the reaction stops, even if other reactants are still present. The other reactants that are not completely used are called **excess reactants**.

Identifying the Limiting Reactant

To identify the limiting reactant, we compare the molar amounts of each reactant to the required mole ratios from the balanced equation.

Example: Limiting Reactant

Consider the reaction between nitrogen and hydrogen to produce ammonia:

$$N_2(g) + 3H_2(g) \rightarrow 2NH_3(g)$$

Suppose you have 10.0 g of nitrogen and 5.0 g of hydrogen. Which is the limiting reactant?

Step 1: Write the balanced equation.

The balanced equation shows 1 mole of N_2 reacts with 3 moles of H_2.

Step 2: Find the molar masses.

N_2 = 28.02 g/mol, H_2 = 2.02 g/mol.

Step 3: Convert the mass of each reactant to moles.

$$\text{Moles of } N_2 = \frac{10.0\,\text{g}}{28.02\,\text{g/mol}} = 0.357\,\text{mol}\,N_2$$

$$\text{Moles of } H_2 = \frac{5.0\,\text{g}}{2.02\,\text{g/mol}} = 2.475\,\text{mol}\,H_2$$

Step 4: Determine the limiting reactant using mole ratios. According to the balanced equation, 1 mole of N_2 reacts with 3 moles of H_2. Thus, 0.357 mol of N_2 would require:

$$0.357\,\text{mol}\,N_2 \times \frac{3\,\text{mol}\,H_2}{1\,\text{mol}\,N_2} = 1.071\,\text{mol}\,H_2$$

Since 1.071 mol of H_2, is needed, and you have 2.475 mol of H_2, there is more than enough hydrogen. Therefore, nitrogen is the limiting reactant, and the reaction will stop when all the nitrogen is consumed.

Calculating the Theoretical Yield

Once the limiting reactant is identified, it can be used to calculate the **theoretical yield**, which is the maximum amount of product that can be formed from the limiting reactant.

Continuing with the ammonia example:

Since 1 mole of N_2 produces 2 moles of NH_3, 0.357 mol of N_2 will produce:

$$0.357\,\text{mol}\,N_2 \times \frac{2\,\text{mol}\,NH_3}{1\,\text{mol}\,N_2} = 0.714\,\text{mol}\,NH_3$$

Convert moles of NH_3 to grams:

$$0.714\,\text{mol}\,NH_3 \times 17.03\,\text{g/mol} = 12.16\,\text{g}\,NH_3$$

Thus, the theoretical yield of ammonia is 12.16 g.

Real-World Applications of Stoichiometry and Limiting Reactants

Industrial Applications

In industrial chemical production, stoichiometry and the concept of limiting reactants are essential for optimizing efficiency and reducing waste. Manufacturers calculate precisely how much of each reactant is needed to minimize excess and cost.

For example, in the production of fertilizers like ammonia (the Haber process), companies optimize the ratio of nitrogen to hydrogen to maximize yield while minimizing energy and resource consumption.

Environmental Chemistry

In environmental chemistry, limiting reactants can influence pollutant formation. For instance, in combustion reactions, insufficient oxygen (the limiting reactant) can lead to incomplete combustion, producing harmful pollutants like carbon monoxide instead of carbon dioxide. Stoichiometric analysis helps in designing processes that reduce these pollutants by ensuring complete combustion.

Mastering stoichiometry and the concept of limiting reactants is fundamental to success in chemistry. These principles not only allow chemists to predict the outcome of reactions but also enable the efficient use of resources in industrial, environmental, and biological processes. By understanding the quantitative relationships in chemical reactions, students can unlock deeper insights into how matter interacts and transforms, paving the way for further exploration and application in the field of chemistry.

Chapter 7: Solutions and Concentrations

Understanding Solutions

Solutions are fundamental to the study of chemistry, playing a crucial role in a wide range of chemical processes, from biological functions to industrial applications. This chapter delves into the nature of solutions, their components, types, and the importance of understanding solutions in both theoretical and practical contexts.

What is a Solution?

A **solution** is a homogeneous mixture of two or more substances. In a solution, one substance (the solute) is dissolved in another substance (the solvent). The uniformity of a solution means that the composition is consistent throughout, allowing for predictable properties and behaviors.

Components of a Solution

Solute: The solute is the substance that is dissolved in the solvent. It can be a solid, liquid, or gas. Common examples include salt in water, sugar in tea, and carbon dioxide in soda.

Solvent: The solvent is the substance in which the solute dissolves. It is typically the component present in the greatest amount. Water is often referred to as the "universal solvent" due to its ability to dissolve many substances.

Types of Solutions

Solutions can be categorized based on the physical states of their solute and solvent:

Aqueous Solutions: Solutions where water is the solvent. Most biochemical and chemical reactions occur in aqueous solutions, making them critical in chemistry.

Solid Solutions: These include alloys (e.g., bronze, steel) where metals are mixed, forming a solid solution.

Gas Solutions: Gaseous mixtures, such as air, where multiple gases (e.g., nitrogen, oxygen, carbon dioxide) are mixed homogeneously.

Liquid Solutions: These can be solutions where both solute and solvent are liquids, like alcohol in water.

How Solutions Form

The process of forming a solution involves the interaction between solute and solvent molecules. When a solute is added to a solvent, intermolecular forces play a key role in the dissolution process:

Breaking Intermolecular Forces: Energy is required to break the intermolecular forces holding the solute molecules together. This is called the lattice energy in solids.

Solvent-Solute Interactions: Once the solute particles are separated, they interact with the solvent molecules. The solvent molecules surround the solute particles, stabilizing them in solution through solvation or hydration in the case of aqueous solutions.

Equilibrium: The process of dissolving continues until an equilibrium is reached, where the rate of solute molecules entering solution equals the rate of solute molecules leaving the solution.

Factors Affecting Solubility

The solubility of a substance refers to its ability to dissolve in a solvent at a given temperature and pressure. Several factors influence solubility:

Nature of the Solute and Solvent: The saying "like dissolves like" summarizes the principle that polar solutes tend to dissolve in polar solvents (e.g., salt in water), while nonpolar solutes dissolve in nonpolar solvents (e.g., oil in hexane).

Temperature: For most solid solutes, solubility increases with temperature. However, for gases, solubility typically decreases with increasing temperature.

Pressure: Pressure has a significant effect on the solubility of gases. According to Henry's Law, the amount of gas that dissolves in a liquid is directly proportional to the partial pressure of that gas above the liquid.

Stirring and Surface Area: Increasing stirring (agitation) or surface area (e.g., by crushing solids) can enhance the rate of dissolution, although they do not change the overall solubility.

Concentration of Solutions

The concentration of a solution is a measure of how much solute is present in a given amount of solvent or solution. Several ways to express concentration include:

Molarity (M): The most common concentration unit, defined as the number of moles of solute per liter of solution.

$$\text{Molarity (M)} = \frac{\text{moles of solute}}{\text{liters of solution}}$$

Molality (m): The number of moles of solute per kilogram of solvent. It is useful for studying colligative properties, especially when temperature changes are involved.

$$\text{Molality (m)} = \frac{\text{moles of solute}}{\text{kilograms of solvent}}$$

Mass Percent: The mass of solute divided by the total mass of the solution, expressed as a percentage.

$$\text{Mass \%} = \left(\frac{\text{mass of solute}}{\text{mass of solution}}\right) \times 100$$

Volume Percent: The volume of solute divided by the total volume of the solution, expressed as a percentage. This is often used for solutions containing liquids.

$$\text{Volume \%} = \left(\frac{\text{volume of solute}}{\text{volume of solution}}\right) \times 100$$

Parts per Million (ppm) and Parts per Billion (ppb): Used to express very dilute concentrations, especially in environmental science and chemistry.

$$\text{ppm} = \left(\frac{\text{mass of solute}}{\text{mass of solution}}\right) \times 10^6$$

$$\text{ppb} = \left(\frac{\text{mass of solute}}{\text{mass of solution}}\right) \times 10^9$$

Importance of Solutions in Chemistry

Understanding solutions is crucial for various reasons:

Chemical Reactions: Many chemical reactions occur in solution, especially in biological systems. Knowledge of solutions aids in predicting reaction behavior, rates, and equilibria.

Biological Processes: Solutions are vital in biological contexts, such as in cellular processes, nutrient transport, and biochemical reactions.

Industrial Applications: Solutions are used in manufacturing processes, pharmaceuticals, food production, and environmental management. Understanding solubility and concentration is key to optimizing these processes.

Environmental Chemistry: Solutions play a role in pollutant behavior, distribution, and degradation in environmental contexts, impacting ecosystems and human health.

Analytical Chemistry: Many analytical techniques, such as titrations and spectrophotometry, rely on the precise understanding of solutions and their concentrations.

Solutions are a cornerstone of chemistry, influencing everything from fundamental research to practical applications. By grasping the concepts of solubility, concentration, and the behavior of solutions, students and practitioners can unlock a deeper understanding of chemical interactions, both in the lab and in the real world. This foundational knowledge sets the stage for more advanced studies and applications across the vast landscape of chemistry.

Solubility and Factors Affecting Solubility

Solubility is a key concept in chemistry that describes the ability of a solute to dissolve in a solvent, resulting in a homogeneous solution. Understanding solubility is critical for predicting how substances behave in various environments, from laboratory experiments to natural systems. This section explores the concept of solubility, the factors that influence it, and its significance in chemistry.

What is Solubility?

Solubility refers to the maximum amount of a solute that can dissolve in a given amount of solvent at a specified temperature and pressure. It is often expressed in units such as grams of solute per 100 grams of solvent or in molarity (moles of solute per liter of solution). The solubility of a substance depends on its chemical nature, the nature of the solvent, temperature, and pressure.

Types of Solubility

Saturated Solutions: A saturated solution is one in which the maximum amount of solute has been dissolved in the solvent at a given temperature and pressure. Any additional solute will not dissolve and will remain undissolved.

Unsaturated Solutions: An unsaturated solution contains less solute than it can theoretically hold at a specific temperature. Additional solute can still dissolve.

Supersaturated Solutions: A supersaturated solution contains more solute than a saturated solution at the same temperature. This state is unstable, and excess solute can precipitate out of the solution if disturbed.

Factors Affecting Solubility

Several key factors influence the solubility of substances:

Nature of the Solute and Solvent:

The solubility of a solute in a solvent is significantly affected by their chemical nature. The principle "like dissolves like" summarizes this concept:

Polar Solvents (e.g., water) are effective at dissolving polar solutes (e.g., salts, sugars) due to the strong intermolecular forces between polar molecules.

Nonpolar Solvents (e.g., hexane) are better at dissolving nonpolar solutes (e.g., oils, fats) because of similar van der Waals forces.

Temperature:

Temperature has a profound effect on solubility, although the impact can vary between solids, liquids, and gases:

Solids: For most solid solutes, solubility increases with temperature. This is due to the increased kinetic energy of molecules, which allows solute particles to overcome intermolecular forces and enter solution more readily.

Gases: For gas solutes, solubility typically decreases with increasing temperature. As temperature rises, gas molecules gain kinetic energy and escape from the solution more easily, leading to lower solubility.

Pressure:

Pressure primarily affects the solubility of gases rather than solids and liquids. According to Henry's Law, the solubility of a gas in a liquid is directly proportional to the partial pressure of that gas above the liquid. Increasing the pressure increases the concentration of gas molecules in the liquid, thus increasing solubility. This principle is applied in various industries, such as carbonated beverages, where carbon dioxide is dissolved under high pressure.

Stirring and Surface Area:

While stirring does not change solubility, it can enhance the rate at which a solute dissolves by increasing contact between solute and solvent molecules. Similarly, increasing the surface area of the solute (e.g., by crushing solid particles) facilitates more effective interactions with the solvent, leading to faster dissolution.

Ionic Strength:

In solutions with multiple ions, the presence of other dissolved ions can affect solubility. This is known as the "common ion effect." For instance, the solubility of calcium sulfate decreases in a solution already containing calcium ions, as the additional calcium ions shift the equilibrium according to Le Chatelier's principle.

pH of the Solution:

The solubility of some compounds is pH-dependent. For example, the solubility of certain metal hydroxides increases in acidic solutions because protons can react with the hydroxide ions, thereby shifting the equilibrium and allowing more of the solid to dissolve.

Measuring Solubility

Solubility is typically determined experimentally by:

Saturation Method: Creating a saturated solution at a specific temperature and measuring the concentration of solute at equilibrium.

Graphical Methods: Plotting solubility data against temperature to observe trends and determine solubility limits at different temperatures.

Using Standard Solutions: Comparing the solubility of unknown solutes with known standards under controlled conditions.

Importance of Solubility in Chemistry

Chemical Reactions: Many chemical reactions occur in solution, making solubility a key factor in reaction kinetics and equilibria. Understanding how solubility affects reaction conditions helps chemists predict product formation and yield.

Biological Processes: Solubility is vital for biochemical reactions, nutrient transport, and metabolic processes within living organisms. For example, the solubility of oxygen in blood is crucial for respiration.

Environmental Science: Solubility influences pollutant behavior in natural water systems. Understanding how pollutants dissolve and interact with water affects environmental impact assessments and remediation strategies.

Pharmaceutical Applications: In drug formulation, solubility determines the bioavailability of medications. Drugs must be soluble to be absorbed effectively in the body, making solubility a critical factor in pharmaceutical development.

Industrial Processes: Many industrial applications, such as crystallization, extraction, and wastewater treatment, rely on solubility principles to optimize processes and achieve desired outcomes.

Solubility is a fundamental concept in chemistry that impacts a wide range of scientific fields, from chemical reactions and biological systems to environmental science and industrial applications. By understanding the factors that affect solubility, chemists can predict how substances behave in solution, which is essential for both theoretical studies and practical applications. Mastery of solubility concepts equips students and professionals with the knowledge needed to navigate the complex interactions of chemicals in various contexts.

Concentration Units

Concentration is a critical concept in chemistry that quantifies the amount of solute present in a given volume or mass of solvent or solution. Understanding and calculating concentration is essential for performing chemical reactions, preparing solutions, and analyzing substances in various contexts. This section explores the different units of concentration, their applications, and how to convert between them.

Common Units of Concentration

Molarity (M)

Molarity is the most commonly used concentration unit in chemistry. It is defined as the number of moles of solute per liter of solution.

$$\text{Molarity (M)} = \frac{\text{moles of solute}}{\text{liters of solution}}$$

Application: Molarity is widely used in stoichiometry, titrations, and reactions in aqueous solutions, providing a clear measure of how concentrated a solution is.

Molality (m)

Molality measures the number of moles of solute per kilogram of solvent. It is particularly useful in situations where temperature changes might affect the volume of the solution.

$$\text{Molality (m)} = \frac{\text{moles of solute}}{\text{kilograms of solvent}}$$

Application: Molality is important for studying colligative properties (such as boiling point elevation and freezing point depression) because it is independent of temperature.

Mass Percent (w/w%)

Mass percent expresses the concentration of a solute as a percentage of the total mass of the solution.

$$\text{Mass \%} = \left(\frac{\text{mass of solute}}{\text{mass of solution}}\right) \times 100$$

Application: Mass percent is commonly used in food labeling, pharmaceuticals, and environmental science to convey the concentration of active ingredients or contaminants.

Volume Percent (v/v%)

Volume percent indicates the volume of solute divided by the total volume of the solution, expressed as a percentage.

$$\text{Volume \%} = \left(\frac{\text{volume of solute}}{\text{volume of solution}}\right) \times 100$$

Application: This unit is often used for liquid solutions, such as alcoholic beverages, where the concentration of the solute (e.g., ethanol) is expressed in relation to the total volume of the liquid.

Parts per Million (ppm)

PPM is used to express very dilute concentrations and is defined as the mass of solute per million parts of solution.

$$\text{ppm} = \left(\frac{\text{mass of solute}}{\text{mass of solution}}\right) \times 10^6$$

Application: PPM is commonly used in environmental science to measure the concentration of pollutants in air, water, and soil.

Parts per Billion (ppb)

Similar to ppm, ppb measures even lower concentrations, defined as the mass of solute per billion parts of solution.

$$\text{ppb} = \left(\frac{\text{mass of solute}}{\text{mass of solution}}\right) \times 10^9$$

Application: PPB is often used in toxicology and environmental monitoring to measure trace contaminants.

Converting Between Concentration Units

Understanding how to convert between different concentration units is essential for practical applications. Here are some common conversion methods:

From Molarity to Mass Percent:

To convert molarity to mass percent, you need the molar mass of the solute and the density of the solution. The formula involves several steps:

Calculate the mass of solute in 1 L of solution (using molarity).

- Calculate the total mass of the solution (mass of solute + mass of solvent).
- Use the mass percent formula.

From Molarity to Molality:

To convert molarity to molality, you need the density of the solution:

- First, calculate the mass of solute in 1 L (moles of solute × molar mass).
- Calculate the mass of the solvent using the density of the solution.
- Then, use the molality formula.
- From Mass Percent to Molarity:

To convert mass percent to molarity:

- Use the mass percent to find the mass of solute in a specific volume of solution.
- Calculate the number of moles of solute.
- Finally, divide the number of moles by the volume of the solution in liters.

Importance of Concentration in Chemistry

Stoichiometry: Concentration units are essential for calculating reactant and product quantities in chemical reactions. Understanding concentrations allows chemists to perform accurate stoichiometric calculations.

Titrations: In titrations, knowing the concentration of solutions is critical for determining the concentration of an unknown solution through reactions with a standard solution.

Pharmaceuticals: Concentration units are vital in drug formulation, ensuring that medications are effective and safe. Understanding dosages in mass percent or molarity can affect therapeutic outcomes.

Environmental Monitoring: Concentration measurements are crucial for assessing pollutant levels in air and water, informing regulatory standards and safety measures.

Quality Control: In industrial applications, monitoring concentrations of substances is necessary to ensure product quality and adherence to safety standards.

Concentration units provide essential tools for quantifying the amount of solute in solutions. By understanding these units and how to convert between them, students and practitioners can navigate a wide range of chemical applications, from laboratory work to industrial processes. Mastery of concentration concepts is fundamental to success in chemistry, facilitating accurate measurements and effective communication of chemical information.

Colligative Properties of Solutions

Colligative properties are an important set of physical properties in solutions that depend solely on the number of solute particles present, rather than the type of particles. These properties arise when a solute is dissolved in a solvent and include key phenomena such as boiling point elevation, freezing point depression, vapor pressure lowering, and osmotic pressure. This section explores the concept of colligative properties in depth, their applications, and the underlying principles governing them.

What are Colligative Properties?

Colligative properties are characteristics of a solution that change when a solute is added to a solvent. Unlike other properties (like color, conductivity, or reactivity), colligative properties depend on the quantity of solute particles (atoms, ions, or molecules), not on the chemical nature of the solute. This means that whether the solute is a sugar molecule, salt ion, or alcohol molecule, the effect on colligative properties will be similar as long as the concentration of particles is the same.

The four primary colligative properties are:

Vapor Pressure Lowering

Boiling Point Elevation

Freezing Point Depression

Osmotic Pressure

Each of these properties results from the interaction between solute particles and the solvent molecules, altering the physical behavior of the solution compared to the pure solvent.

1. Vapor Pressure Lowering

Vapor pressure refers to the pressure exerted by the vapor of a liquid in equilibrium with its liquid phase at a given temperature. When a non-volatile solute (one that doesn't evaporate easily) is dissolved in a solvent, it reduces the number of solvent molecules at the surface that can escape into the vapor phase. This leads to a reduction in the vapor pressure of the solution compared to the pure solvent.

Raoult's Law governs this phenomenon, stating that the vapor pressure of the solvent in a solution is proportional to the mole fraction of the solvent. Mathematically:

$$P_{\text{solution}} = X_{\text{solvent}} P_{\text{solvent}}$$

Where:

P_{solution} is the vapor pressure of the solution,

X_{solvent} is the mole fraction of the solvent,

P_{solvent} is the vapor pressure of the pure solvent.

Application: Vapor pressure lowering is critical in processes such as distillation, where understanding the behavior of vapor pressure is essential for separating components of a mixture.

2. Boiling Point Elevation

Boiling point elevation is a direct result of vapor pressure lowering. As the vapor pressure of a solution decreases with the addition of a solute, a higher temperature is required for the solution's vapor pressure to equal atmospheric pressure, which is the condition for boiling.

The increase in boiling point (ΔT_b) is proportional to the molality (m) of the solute and is given by the equation:

$$\Delta T_b = iK_b m$$

Where:

ΔT_f is the freezing point depression,

i is the van't Hoff factor,

K_f is the cryoscopic constant of the solvent,

m is the molality of the solution.

Example: This effect is why adding salt to icy roads helps to melt the ice. The salt lowers the freezing point of water, allowing the ice to melt at a lower temperature.

Application: Freezing point depression is widely used in real-world applications such as de-icing, making ice cream, and in the development of antifreeze solutions that prevent water from freezing in extreme temperatures.

4. Osmotic Pressure

Osmosis is the movement of solvent molecules through a semi-permeable membrane from a region of lower solute concentration to a region of higher solute concentration. **Osmotic pressure** is the pressure required to stop this flow of solvent molecules and maintain equilibrium. It is directly proportional to the concentration of solute particles in the solution.

Osmotic pressure (Π) can be calculated using the formula:

$$\Pi = iMRT$$

Where:

Π is the osmotic pressure,

i is the van't Hoff factor,

M is the molarity of the solution,

R is the gas constant,

T is the temperature in Kelvin.

Example: Osmotic pressure is crucial in biological systems. For instance, cells rely on osmotic pressure to regulate the movement of water and maintain cell structure.

Application: Osmosis and osmotic pressure are used in water purification processes such as reverse osmosis, where pressure is applied to force water through a membrane, leaving solute particles behind. This process is vital for desalination plants that convert seawater into freshwater.

The van't Hoff Factor (i)

A key concept in colligative properties is the **van't Hoff factor** (i), which represents the number of particles into which a solute dissociates in solution. For example:

Non-electrolytes (e.g., sugar) have an i value of 1 because they do not dissociate into ions.

Electrolytes (e.g., NaCl) dissociate into ions, so i is greater than 1. For NaCl, $i = 2$ because it dissociates into two ions: Na^+ and Cl^-.

Separating Mixtures

Separating mixtures is a fundamental process in both chemistry and everyday life, allowing the isolation of individual components from a mixture. Unlike chemical compounds, where elements are chemically bonded, mixtures involve substances that are physically combined and can be separated by physical means without altering their chemical identities. Understanding the techniques for separating mixtures is crucial in laboratory work, industrial processes, and even daily activities like purifying water or separating recyclables.

Mixtures can be classified into **homogeneous** and **heterogeneous** mixtures. The method used to separate them often depends on their physical properties such as particle size, solubility, boiling point, or magnetic properties. This section discusses the most common methods for separating mixtures and their practical applications.

Types of Mixtures

Homogeneous Mixtures: These mixtures have uniform composition throughout, meaning the components are indistinguishable (e.g., saltwater, air). Techniques such as distillation and chromatography are often used to separate the components of homogeneous mixtures.

Heterogeneous Mixtures: These mixtures have visibly different components and phases (e.g., sand and water, oil and vinegar). Methods like filtration, decantation, and centrifugation are effective for separating heterogeneous mixtures.

Methods for Separating Mixtures

1. Filtration

Filtration is a method used to separate an insoluble solid from a liquid in a heterogeneous mixture. It works by passing the mixture through a filter, which allows the liquid (filtrate) to pass through while retaining the solid (residue) on the filter.

Example: A common use of filtration is in coffee brewing, where coffee grounds are separated from the liquid coffee using a paper filter.

Application: Filtration is widely used in water purification, air filtration systems, and in laboratories to isolate precipitates formed during chemical reactions.

Types of Filtration:

Gravity Filtration: Used when a solution needs to be separated under the influence of gravity.

Vacuum Filtration: More efficient than gravity filtration, this method uses a vacuum to pull the liquid through the filter faster.

2. Distillation

Distillation is a separation technique that relies on the difference in boiling points of the substances in a homogeneous mixture. The process involves heating the mixture until one component evaporates, then condensing and collecting the vapor in a separate container.

Example: Separating alcohol from water, where alcohol boils at a lower temperature than water.

Application: Distillation is crucial in industries such as petroleum refining (fractional distillation) and the production of alcoholic beverages. It is also used to purify liquids, such as producing distilled water.

Types of Distillation:

Simple Distillation: Used when separating two liquids with significantly different boiling points or separating a liquid from non-volatile impurities.

Fractional Distillation: Used when separating a mixture of liquids with closer boiling points. This method is widely used in oil refineries to separate crude oil into its components like gasoline, diesel, and kerosene.

3. Chromatography

Chromatography is a powerful technique for separating mixtures based on the differential movement of solute components through a medium. The mixture is carried by a mobile phase (liquid or gas) through a stationary phase (solid or liquid), and different components travel at different speeds, leading to separation.

Example: Paper chromatography is often used to separate pigments in ink.

Application: Chromatography is essential in pharmaceuticals, biochemistry, and environmental testing to analyze complex mixtures. Gas chromatography (GC) and high-performance liquid chromatography (HPLC) are widely used for separating and analyzing volatile and non-volatile compounds, respectively.

Types of Chromatography:

Paper Chromatography: Used for separating pigments or dyes.

Thin-Layer Chromatography (TLC): Used for qualitative analysis of organic compounds.

Gas Chromatography (GC): Used for separating and analyzing volatile substances.

High-Performance Liquid Chromatography (HPLC): Used for separating and analyzing compounds that are not easily volatilized.

4. Evaporation

Evaporation is a process where the solvent in a solution is heated until it evaporates, leaving behind the solute. This method is effective for separating a soluble solid from a liquid.

Example: Evaporating saltwater to obtain salt crystals.

Application: Evaporation is commonly used in industries like salt production, where large amounts of seawater are evaporated to produce salt. It is also used in the food industry, such as in the production of concentrated fruit juices.

5. Centrifugation

Centrifugation is a technique that uses centrifugal force to separate components of a heterogeneous mixture based on their densities. The mixture is spun at high speeds, causing denser particles to settle at the bottom, while less dense components remain on top.

Example: Centrifugation is used in blood tests to separate blood cells from plasma.

Application: This method is widely used in biochemistry, clinical laboratories, and wastewater treatment to separate suspended particles from liquids.

6. Decantation

Decantation involves carefully pouring off a liquid to separate it from a solid or another liquid that has settled at the bottom. This method is suitable when the solid particles are large and settle quickly.

Example: Decanting wine to separate it from sediment.

Application: Decantation is used in industries such as mining, where minerals are separated from water or other liquids.

7. Magnetic Separation

Magnetic separation is used to separate magnetic materials from a mixture. If one of the components of a mixture is magnetic, a magnet can attract it, leaving the non-magnetic substances behind.

Example: Separating iron filings from a mixture of sand and iron.

Application: This method is used in recycling industries to separate metal from non-metal waste and in mining to extract magnetic minerals.

8. Crystallization

Crystallization is a method where a solution is cooled or evaporated, allowing the dissolved solute to form crystals. This method is used to purify solid substances.

Example: Crystallization is used to purify sugar from sugarcane juice.

Application: Crystallization is widely used in the pharmaceutical industry to purify chemical compounds and in the production of high-purity chemicals like salts and sugars.

9. Sublimation

Sublimation is the process by which a solid changes directly into a gas without passing through the liquid phase. This method is used to separate substances that can sublime from those that cannot.

Example: Sublimation of iodine or dry ice (solid carbon dioxide).

Application: Sublimation is used in the purification of compounds, especially in the electronics industry to produce high-purity substances.

Importance of Separating Mixtures

Environmental Applications: Separation techniques are critical in environmental science for water purification, air quality testing, and pollutant removal. Processes like filtration and distillation help provide clean drinking water and reduce environmental contaminants.

Medical and Pharmaceutical Fields: In medicine, separation methods like centrifugation and chromatography are used to analyze blood samples, purify drugs, and develop new medications.

Industrial Processes: Industries rely heavily on separation techniques for manufacturing, refining raw materials, and producing goods. For example, oil refineries use fractional distillation to separate crude oil into various fuel types.

Scientific Research: In laboratories, scientists regularly use separation techniques to isolate and analyze chemical components, enabling discoveries in fields such as biochemistry, pharmacology, and material science.

Separating mixtures is a foundational concept in chemistry and has vast applications in both scientific research and everyday life. Understanding the principles behind each technique and knowing when to apply them is critical in solving practical problems, from purifying water to developing new medicines. These methods, though physical, are crucial tools that support more complex chemical processes, making them indispensable in modern science and industry.

Chapter 8: Acids and Bases

Properties of Acids and Bases

Acids and bases are two fundamental categories of chemical substances that play vital roles in various chemical reactions, industrial processes, and biological systems. Understanding their properties is essential to recognizing their behavior in reactions and their practical applications in everyday life. Below is a detailed discussion of the properties of acids and bases, focusing on their physical, chemical, and observable characteristics.

Properties of Acids

Acids are substances that donate protons (H^+ ions) or accept electrons during chemical reactions. The presence of hydrogen ions in an aqueous solution is what gives acids their characteristic properties. These can vary from strong acids, like hydrochloric acid (HCl), to weak acids, such as acetic acid (CH_3COOH).

1. Taste

Sour Taste: One of the most well-known properties of acids is their sour taste. This is observable in foods like lemons, vinegar, and other sour-tasting substances that contain acids such as citric acid or acetic acid. However, it's important to note that tasting chemicals in a laboratory setting is extremely dangerous and should never be attempted.

Examples: Lemon juice contains citric acid, and vinegar contains acetic acid, both of which give these substances their distinct sour taste.

2. pH Value

Low pH (Below 7): Acids have a pH value of less than 7 on the pH scale, which ranges from 0 to 14. Strong acids like sulfuric acid (H_2SO_4) have pH values close to 0, while weaker acids like acetic acid have pH values closer to 7. The lower the pH, the stronger the acid.

pH Indicator Reaction: Acids turn blue litmus paper red. This is one of the standard tests used in laboratories to identify an acidic substance.

3. Reactivity with Metals

Reaction with Metals to Produce Hydrogen Gas: Acids typically react with metals, particularly active metals like magnesium (Mg), zinc (Zn), and iron (Fe), to produce hydrogen gas (H_2) and a salt. This reaction is commonly observed in strong acids like hydrochloric acid.

Example: Zinc reacts with hydrochloric acid to produce zinc chloride ($ZnCl_2$) and hydrogen gas:

$$Zn(s) + 2HCl(aq) \rightarrow ZnCl_2(aq) + H_2(g)$$

4. Electrical Conductivity

Electrolytic Behavior: Acids are good conductors of electricity when dissolved in water because they dissociate into ions (H^+ ions and anions). The free ions in solution allow the passage of electrical current, making acids electrolytes.

Example: Sulfuric acid in solution is a strong electrolyte and is used in lead-acid batteries to conduct electricity.

5. Corrosive Nature

Corrosiveness: Acids, especially strong acids, can be highly corrosive. They can cause damage to skin, metals, and organic materials. For instance, sulfuric acid and hydrochloric acid are highly corrosive and can cause chemical burns upon contact with skin.

Example: Industrial acids are handled with care, as concentrated sulfuric acid can burn through metals, plastics, and human tissue.

6. Neutralization Reaction

Reaction with Bases: Acids neutralize bases to form water and a salt. This reaction is called a **neutralization reaction** and is essential in various applications, such as in antacids to relieve heartburn.

Example: The reaction between hydrochloric acid and sodium hydroxide (a base) forms sodium chloride (salt) and water:

$$HCl(aq) + NaOH(aq) \rightarrow NaCl(aq) + H_2O(l)$$

7. Reaction with Carbonates

Carbon Dioxide Gas Production: Acids react with carbonates and bicarbonates to produce carbon dioxide (CO_2), water, and a salt. This is often observed in effervescent reactions.

Example: When hydrochloric acid reacts with sodium bicarbonate (baking soda), the products are sodium chloride, carbon dioxide, and water:

$$HCl(aq) + NaHCO_3(s) \rightarrow NaCl(aq) + H_2O(l) + CO_2(g)$$

Properties of Bases

Bases, also known as alkalis when dissolved in water, are substances that can accept protons or donate electron pairs in reactions. They produce hydroxide ions (OH^-) when dissolved in water. Common bases include sodium hydroxide (NaOH) and ammonia (NH_3).

1. Taste

Bitter Taste: Bases typically have a bitter taste, although tasting them is highly discouraged due to their corrosive properties. This property is evident in substances like baking soda (sodium bicarbonate).

Example: Many cleaning products that contain bases, such as soaps and detergents, have a slight bitterness if ingested.

2. pH Value

High pH (Above 7): Bases have pH values greater than 7, with strong bases like sodium hydroxide having pH values close to 14, and weak bases like ammonia having pH values closer to 7. The higher the pH, the stronger the base.

pH Indicator Reaction: Bases turn red litmus paper blue. This is a standard test to identify a basic solution in laboratories.

3. Slippery or Soapy Feel

Texture: Aqueous solutions of bases often feel slippery or soapy to the touch. This is because they react with oils on the skin to form soap-like substances, making the surface feel slippery.

Example: Sodium hydroxide (lye) in soap-making reacts with fats to form soap, which gives soap its slippery feel.

4. Reactivity with Acids

Neutralization: Like acids, bases undergo neutralization reactions, but in this case, they react with acids to produce water and salt. This reaction is exothermic, meaning it releases heat.

Example: Calcium hydroxide (lime) reacts with carbonic acid in water to neutralize acidity, forming calcium carbonate and water.

5. Electrical Conductivity

Good Conductors: Bases conduct electricity in solution due to the presence of hydroxide ions (OH^-), making them electrolytes. Strong bases like sodium hydroxide dissociate completely in water, allowing electricity to flow through the solution efficiently.

Example: Sodium hydroxide in water is a strong conductor of electricity and is often used in electrolysis processes.

6. Corrosive Nature

Causticity: Strong bases like sodium hydroxide and potassium hydroxide are highly caustic and can cause severe chemical burns upon contact with skin. Bases can corrode organic materials and metals, though they are more commonly associated with breaking down proteins and fats.

Example: Lye (sodium hydroxide) is extremely caustic and is used in industrial cleaning agents, but it must be handled carefully due to its potential to burn skin.

7. Reaction with Oils and Fats

Saponification: Bases react with fats and oils in a process called **saponification** to produce soap. This reaction is important in both industrial and homemade soap production.

Example: Sodium hydroxide reacts with animal fats to form soap and glycerol:

$$\text{Fat} + NaOH \rightarrow \text{Soap} + \text{Glycerol}$$

8. Ammonia and Weak Bases

Weak Base Properties: Ammonia (NH_3) is a common weak base that does not fully dissociate in water. However, it still increases the pH of the solution and is used widely in cleaning products and fertilizers.

Example: Ammonia in water forms ammonium hydroxide, which is mildly alkaline and used for household cleaning.

The properties of acids and bases define their unique roles in chemistry and their applications in everyday life. Acids are sour, conduct electricity, and react with metals and bases, while bases are bitter, slippery, and neutralize acids. Their behavior is dictated by their pH and ability to donate or accept protons. These properties are not only essential in the laboratory and industry but also in natural processes, from digestion in the human stomach (acids) to the production of soap (bases). Understanding the basic characteristics of acids and bases helps explain their extensive use in food, cleaning agents, medicine, and industrial processes.

Theories of Acids and Bases

Over time, scientists have developed several theories to explain the behavior of acids and bases. These theories help describe the nature of acids and bases, how they interact with one another, and how they behave in different chemical environments. The three most important and widely recognized theories are:

Arrhenius Theory

Brønsted-Lowry Theory

Lewis Theory

Each of these theories has contributed to a deeper understanding of acid-base chemistry, and they provide distinct frameworks that apply to various chemical contexts. Let's explore each theory in detail.

1. Arrhenius Theory of Acids and Bases

The **Arrhenius theory** is one of the earliest and simplest models for describing acids and bases. Proposed by the Swedish chemist Svante Arrhenius in 1884, it focuses primarily on the behavior of acids and bases in aqueous solutions.

Definition of Acids and Bases:

Arrhenius Acid: An Arrhenius acid is a substance that, when dissolved in water, increases the concentration of hydrogen ions (H^+) or hydronium ions (H_3O^+) in solution.

Example: Hydrochloric acid (HCl) dissociates in water to produce H^+ ions

$$HCl \rightarrow H^+ + Cl^-$$

The H^+ ions combine with water molecules to form H_3O^+:

$$H^+ + H_2O \rightarrow H_3O^+$$

Arrhenius Base: An Arrhenius base is a substance that, when dissolved in water, increases the concentration of hydroxide ions (OH^-).

Example: Sodium hydroxide (NaOH) dissociates in water to produce OH^- ions:

$$NaOH \rightarrow Na^+ + OH^-$$

Neutralization Reaction:

According to the Arrhenius theory, acids and bases react in aqueous solutions to form water and a salt. This process is called **neutralization**:

$$H^+ + OH^- \rightarrow H_2O$$

Example: When hydrochloric acid (HCl) reacts with sodium hydroxide (NaOH), they neutralize each other to form sodium chloride (NaCl) and water:

$$HCl + NaOH \rightarrow NaCl + H_2O$$

Limitations of the Arrhenius Theory:

Only Applies to Aqueous Solutions: This theory only considers acids and bases that dissolve in water. It cannot explain the behavior of acids and bases in non-aqueous solvents.

Hydroxide Limitation: The Arrhenius definition of bases is limited to compounds that produce OH^- ions. Substances like ammonia (NH_3), which can act as a base but do not release OH^- ions directly, do not fit into the Arrhenius model.

2. Brønsted-Lowry Theory of Acids and Bases

The **Brønsted-Lowry theory**, developed independently by Johannes Brønsted and Thomas Lowry in 1923, is a more comprehensive and flexible theory that applies to a broader range of acid-base reactions, not just those occurring in water.

Definition of Acids and Bases:

Brønsted-Lowry Acid: A Brønsted-Lowry acid is a substance that donates a proton (H^+) in a chemical reaction.

Example: Hydrochloric acid (HCl) donates a proton to water:

$$HCl + H_2O \rightarrow H_3O^+ + Cl^-$$

Brønsted-Lowry Base: A Brønsted-Lowry base is a substance that accepts a proton (H^+) during a chemical reaction.

Example: Ammonia (NH_3) acts as a base by accepting a proton from water:

$$NH_3 + H_2O \rightarrow NH_4^+ + OH^-$$

In this reaction, ammonia accepts a proton (H^+) from water to form the ammonium ion (NH_4^+).

Conjugate Acid-Base Pairs:

A key feature of the Brønsted-Lowry theory is the concept of **conjugate acid-base pairs**. When an acid donates a proton, it forms a conjugate base, and when a base accepts a proton, it forms a conjugate acid.

Example: In the reaction of HCl and water:

$$HCl + H_2O \rightarrow H_3O^+ + Cl^-$$

HCl is the acid, and its conjugate base is Cl^-.

H_2O is the base, and its conjugate acid is H_3O^+.

Versatility of the Brønsted-Lowry Theory:

Applicable in Non-Aqueous Solvents: This theory applies not only to aqueous solutions but also to reactions in gases, non-aqueous solvents, and other phases.

Explains a Wider Range of Reactions: Substances that don't fit into the Arrhenius definition, such as ammonia, can be explained as bases under the Brønsted-Lowry theory.

Example of Non-Aqueous Reaction:

Consider the reaction between ammonia (NH_3) and hydrogen chloride gas (HCl):

$$NH_3 + HCl \rightarrow NH_4Cl$$

Here, ammonia acts as a base by accepting a proton from HCl, forming ammonium chloride (NH_4Cl).

Limitations of the Brønsted-Lowry Theory:

Does Not Address Electron-Pair Donation: While it expands the concept of acids and bases beyond water-based reactions, the Brønsted-Lowry theory does not account for reactions where bases donate electron pairs instead of protons, such as Lewis acid-base reactions.

3. Lewis Theory of Acids and Bases

The **Lewis theory**, introduced by Gilbert N. Lewis in 1923, provides an even broader definition of acids and bases by focusing on electron pairs rather than protons. This theory is particularly useful in explaining reactions that do not involve hydrogen ions.

Definition of Acids and Bases:

Lewis Acid: A Lewis acid is a substance that can accept an electron pair to form a new chemical bond.

Example: Boron trifluoride (BF_3) is a Lewis acid because it can accept an electron pair from a base:

$$BF_3 + NH_3 \rightarrow BF_3NH_3$$

- In this reaction, boron trifluoride accepts an electron pair from ammonia, forming a bond between the two molecules.

Lewis Base: A Lewis base is a substance that donates an electron pair to form a bond.

Example: Ammonia (NH_3) acts as a Lewis base because it donates its lone pair of electrons to boron trifluoride (BF_3).

Importance of the Lewis Theory:

Electron-Pair Focus: The Lewis theory emphasizes the role of electron pairs in acid-base reactions. A Lewis base donates an electron pair, while a Lewis acid accepts it. This definition applies to a wide range of chemical reactions, not just those involving hydrogen ions.

Broad Range of Reactions: Lewis acids include substances that do not contain hydrogen, such as metal cations (e.g., Al^{3+}), and molecules like carbon dioxide (CO_2), which can act as a Lewis acid when reacting with bases like hydroxide ions (OH^-).

Examples of Lewis Acid-Base Reactions:

Complex Formation: Metal cations like iron(III) (Fe^{3+}) act as Lewis acids by accepting electron pairs from Lewis bases such as cyanide ions (CN^-) to form coordination complexes:

$$Fe^{3+} + 6CN^- \rightarrow [Fe(CN)_6]^{3-}$$

Organic Chemistry: In many organic reactions, such as electrophilic aromatic substitution, the electrophile (e.g., a carbocation) acts as a Lewis acid by accepting electrons from the aromatic ring (which acts as a Lewis base).

Strengths of the Lewis Theory:

Covers a Wide Range of Reactions: The Lewis theory explains reactions that are not covered by the Arrhenius or Brønsted-Lowry models, especially in the fields of coordination chemistry, organic chemistry, and industrial catalysis.

No Need for Protons: Unlike the other theories, the Lewis theory does not require acids and bases to involve hydrogen ions. This makes it more universally applicable.

Limitations of the Lewis Theory:

Less Specific to Acid-Base Chemistry: The Lewis theory is broader but lacks some of the specificity seen in the Brønsted-Lowry and Arrhenius models, particularly when discussing reactions in aqueous solutions.

Harder to Identify Acid-Base Strength: The strength of Lewis acids and bases is not as straightforward to determine as it is with pH-based models, making it more difficult to quantify their relative strengths in some cases.

pH and pOH: Measuring Acidity and Basicity

The concepts of pH and pOH are essential for understanding the acidity and basicity of solutions. They provide a quantitative measure of how acidic or basic a substance is, which is critical in a variety of chemical, biological, environmental, and industrial applications. These concepts are closely tied to the concentrations of hydrogen ions (H^+) and hydroxide ions (OH^-) in solutions and are governed by the dissociation of water.

Let's explore the definitions, calculations, and importance of pH and pOH, as well as their relationship with each other.

1. pH: Measuring Acidity

The **pH** scale is a logarithmic scale that quantifies the concentration of hydrogen ions (H^+) or, more precisely, hydronium ions (H_3O^+) in a solution. The pH scale ranges from 0 to 14, where:

- **pH < 7**: Indicates an acidic solution.
- **pH = 7**: Indicates a neutral solution.
- **pH > 7**: Indicates a basic (alkaline) solution.

Definition of pH:

pH is defined mathematically as the negative logarithm of the hydrogen ion concentration $[H^+]$ in moles per liter (M):

$$\mathrm{pH} = -\log[H^+]$$

This equation means that as the concentration of hydrogen ions increases, the pH value decreases, making the solution more acidic. Conversely, as the concentration of H^+ decreases, the pH increases, indicating a more basic solution.

Examples of pH Values:

- **Strongly Acidic Solution**: Hydrochloric acid (HCl) in water, which dissociates completely, results in a high concentration of H^+ ions, giving a pH close to 1.
- **Neutral Solution**: Pure water has a very low concentration of H^+ ions (1×10^{-7} M), resulting in a neutral pH of 7.
- **Strongly Basic Solution**: Sodium hydroxide (NaOH) dissociates to produce OH^- ions, which decrease the H^+ concentration, resulting in a pH close to 13.

Importance of pH:

- **Biological Systems**: The pH of blood (around 7.4) is tightly regulated. Any significant deviation can lead to severe health consequences.
- **Industrial Applications**: pH control is crucial in chemical manufacturing, food processing, water treatment, and pharmaceuticals.
- **Environmental Concerns**: The pH of natural water bodies affects aquatic life. Acid rain, for example, lowers the pH of lakes and streams, harming wildlife.

2. pOH: Measuring Basicity

While pH measures the concentration of hydrogen ions, **pOH** measures the concentration of hydroxide ions (OH^-). Just like pH, pOH is also a logarithmic scale that indicates

the basicity of a solution, with the following relationships:

- **pOH < 7**: Indicates a basic (alkaline) solution.
- **pOH = 7**: Indicates a neutral solution.
- **pOH > 7**: Indicates an acidic solution.

Definition of pOH:

pOH is defined mathematically as the negative logarithm of the hydroxide ion concentration $[OH^-]$

$$\mathrm{pOH} = -\log[OH^-]$$

This equation means that as the concentration of hydroxide ions increases, the pOH value decreases, indicating a more basic solution. Conversely, as the concentration of OH^- decreases, the pOH increases, indicating a more acidic solution.

Examples of pOH Values:

Strongly Basic Solution: Sodium hydroxide (NaOH) in water, which dissociates completely to form a high concentration of OH^- ions, gives a low pOH (close to 1).

Neutral Solution: Pure water has equal concentrations of H^+ and OH^- ions (1×10^{-7} M each), giving both a pH and pOH of 7.

Strongly Acidic Solution: Hydrochloric acid (HCl) in water decreases the OH^- concentration, leading to a high pOH (close to 13).

3. Relationship Between pH and pOH

The relationship between pH and pOH is governed by the ionization of water. Water dissociates into hydrogen ions (H^+) and hydroxide ions (OH^-) according to the equation:

$$H_2O \rightleftharpoons H^+ + OH^-$$

$$H_2O \rightleftharpoons H^+ + OH^-$$

At 25°C (room temperature), the concentration of both H^+ and OH^- in pure water is 1×10^{-7} M, and the **ion product of water** (K_w) is:

$$K_w = [H^+][OH^-] = 1 \times 10^{-14}$$

This leads to the important relationship between pH and pOH:

$$\text{pH} + \text{pOH} = 14$$

This equation holds true for all aqueous solutions at 25°C. Therefore, if you know either the pH or the pOH of a solution, you can easily calculate the other.

Example Calculations:

If the pH of a solution is 3 (acidic), the pOH can be calculated as:

$$\text{pOH} = 14 - 3 = 11$$

If the pOH of a solution is 4 (basic), the pH is:

$$\text{pH} = 14 - 4 = 10$$

4. Importance of pH and pOH in Chemistry

Titration:

pH measurements are crucial in **acid-base titrations**, where the concentration of an unknown solution is determined by reacting it with a standard solution. By monitoring the pH change during the titration process, chemists can identify the equivalence point, where the amount of acid equals the amount of base.

Buffer Solutions:

Buffers are solutions that resist changes in pH when small amounts of acid or base are added. They are vital in biological systems (e.g., blood) and industrial processes, where maintaining a specific pH range is critical.

Industrial and Environmental Applications:

Water Treatment: Controlling the pH is essential in processes like wastewater treatment, ensuring that water is safe for consumption or release into the environment.

Agriculture: Soil pH affects nutrient availability and plant health. Farmers often monitor and adjust soil pH for optimal crop growth.

Environmental Monitoring: Acid rain, caused by pollutants like sulfur dioxide, can lower the pH of lakes and rivers, harming aquatic life. Regular pH monitoring helps track the health of ecosystems.

The concepts of pH and pOH are indispensable for understanding the acidity and basicity of solutions. The pH scale provides an easy way to measure how acidic or basic a solution is, while the pOH scale gives insight into the concentration of hydroxide ions. Both are interconnected through the ion product of water, allowing chemists to predict and control the behavior of solutions in various contexts, from biological systems to industrial applications. Understanding these concepts is foundational to mastering acid-base chemistry.

Acid-Base Reactions and Neutralization

Acid-base reactions are fundamental chemical processes that play a vital role in many natural, industrial, and laboratory settings. They occur when acids and bases interact, resulting in the formation of water and a salt. This process is often referred to as **neutralization**, as it typically neutralizes the acidic and basic properties of the reactants.

To fully understand acid-base reactions and neutralization, it's important to explore the principles behind acids and bases, the reaction mechanism, and the applications of these reactions.

1. Overview of Acid-Base Reactions

In an **acid-base reaction**, an acid donates a proton (H^+) to a base, which accepts the proton. This proton transfer is the key mechanism behind all acid-base reactions. The general form of an acid-base reaction can be expressed as:

$$\text{Acid} + \text{Base} \rightarrow \text{Salt} + \text{Water}$$

Example of a Simple Acid-Base Reaction:

Consider the reaction between hydrochloric acid (HCl) and sodium hydroxide (NaOH):

$$\mathrm{HCl} + \mathrm{NaOH} \rightarrow \mathrm{NaCl} + \mathrm{H_2O}$$

In this reaction:

HCl (hydrochloric acid) acts as the acid, donating a proton (H^+).

NaOH (sodium hydroxide) acts as the base, accepting the proton.

The result is the formation of **NaCl** (sodium chloride, a salt) and water (**H_2O**), effectively neutralizing both the acidic and basic components.

2. Neutralization Reaction

A **neutralization reaction** occurs when an acid and a base react to form water and a salt, with the pH of the solution moving closer to neutral (pH 7). The hydrogen ion (H^+) from the acid combines with the hydroxide ion (OH^-) from the base to form water:

$$H^+ + OH^- \rightarrow H_2O$$

This is a classic example of a neutralization process. While water is always a product, the salt formed depends on the specific acid and base involved.

Example of a Strong Acid-Strong Base Neutralization:

When **sulfuric acid (H_2SO_4)** reacts with **potassium hydroxide (KOH)**, the reaction is as follows:

$$H_2SO_4 + 2KOH \rightarrow K_2SO_4 + 2H_2O$$

Here, potassium sulfate (K_2SO_4), a salt, and water are produced, with the acidity of the sulfuric acid and the basicity of the potassium hydroxide both neutralized.

3. Types of Acid-Base Reactions

Acid-base reactions can be classified based on the strengths of the acids and bases involved, as well as the specific acid-base theory being applied (Arrhenius, Brønsted-Lowry, or Lewis). Each of these theories offers a different way of understanding acid-base behavior, but all can describe acid-base reactions.

a. Strong Acid and Strong Base

When a strong acid reacts with a strong base, the reaction goes to completion, meaning all the acid and base are consumed to form water and salt. The resulting solution is often neutral, with a pH close to 7, because both the acid and base fully dissociate.

Example:

$$CH_3COOH + NH_3 \rightarrow CH_3COONH_4$$

Acetic acid reacts with ammonia to form ammonium acetate (CH_3COONH_4). The resulting solution could be neutral, acidic, or basic, depending on the concentrations and relative strengths of the acid and base.

4. Applications of Neutralization Reactions

Neutralization reactions have widespread applications in many fields, including biology, environmental science, industry, and everyday life.

a. Medical Applications

Antacids: Neutralization reactions are used to relieve heartburn and indigestion. Antacids contain weak bases like magnesium hydroxide or calcium carbonate that react with excess stomach acid (HCl), neutralizing it and alleviating discomfort.

b. Environmental Applications

Acid Rain Mitigation: Acid rain, primarily caused by sulfur dioxide (SO_2) and nitrogen oxides (NO□)in the atmosphere, can lower the pH of soils and water bodies, harming ecosystems. Neutralization reactions using basic substances like calcium carbonate ($CaCO_3$) or lime are employed to neutralize acidic soils and water.

Water Treatment: Neutralization reactions are crucial in treating acidic wastewater or industrial effluents before they are released into the environment.

c. Industrial Applications

Manufacturing: Neutralization is used in the production of various salts, which are key ingredients in many industrial processes. For example, sodium chloride, calcium sulfate, and ammonium nitrate are produced via neutralization reactions.

Agriculture: Farmers often use lime (calcium oxide) to neutralize acidic soil, improving the availability of nutrients to plants.

d. Household Uses

Cleaning Products: Many household cleaning products rely on neutralization. For instance, vinegar (acetic acid) is used to neutralize alkaline residues from soap or hard water deposits.

Baking: In baking, baking soda (sodium bicarbonate) reacts with acids like vinegar or lemon juice to produce carbon dioxide gas, which helps dough rise.

5. Buffer Systems and Neutralization

While neutralization reactions tend to drive the pH of a solution toward 7, **buffer systems** play an important role in maintaining a stable pH. Buffers are solutions that can resist changes in pH when small amounts of acid or base are added. They typically consist of a weak acid and its conjugate base, or a weak base and its conjugate acid. Buffers are critical in biological systems, especially in maintaining the pH of blood.

Acid-base reactions and neutralization are foundational concepts in chemistry with broad applications in daily life, industry, environmental science, and biology. Through the transfer of protons between acids and bases, these reactions lead to the formation of water and salts, often neutralizing the solution's pH in the process. Understanding the types of acid-base reactions, the products formed, and their practical applications provides valuable insight into both theoretical chemistry and real-world processes.

Buffer Solutions and their Applications

Buffer solutions are essential components in many chemical and biological processes. They help maintain a stable pH in a solution, even when acids or bases are added. This stability is critical in various applications, from biological systems to industrial processes. Understanding buffer solutions involves exploring their composition, how they function, and their importance in real-world scenarios.

1. What is a Buffer Solution?

A **buffer solution** is a system that resists changes in pH upon the addition of small amounts of acids or bases. Buffers typically consist of a **weak acid and its conjugate base** or a **weak base and its conjugate acid**.

Examples of Buffer Systems:

Acidic Buffer: A mixture of acetic acid (CH_3COOH) and sodium acetate (CH_3COONa).

Basic Buffer: A mixture of ammonia (NH_3) and ammonium chloride (NH_4Cl).

2. How Buffers Work

The effectiveness of a buffer depends on its components and their concentrations. Here's how buffers maintain pH:

a. Weak Acid and Conjugate Base:

When a small amount of strong acid (H^+) is added to the buffer:

The weak acid (HA) in the buffer can react with the added hydrogen ions:

$$\mathrm{HA} + \mathrm{H}^+ \rightarrow \mathrm{A}^- + \mathrm{H_2O}$$

This reaction consumes the additional H^+ ions, minimizing the change in pH.

When a small amount of strong base (OH^-) is added:

The conjugate base (A^-) can react with the hydroxide ions:

$$\mathrm{A}^- + \mathrm{OH}^- \rightarrow \mathrm{HA} + \mathrm{H_2O}$$

This reaction neutralizes the OH^- ions, again minimizing the change in pH.

b. Weak Base and Conjugate Acid:

For a basic buffer, the process is similar:

If H^+ is added, the conjugate base (B) reacts to neutralize it:

$$\mathrm{B} + \mathrm{H}^+ \rightarrow \mathrm{BH}^+$$

If OH^- is added, the weak base can react with it:

$$\mathrm{BH}^+ + \mathrm{OH}^- \rightarrow \mathrm{B} + \mathrm{H_2O}$$

3. Henderson-Hasselbalch Equation

The effectiveness of a buffer can be quantified using the **Henderson-Hasselbalch equation**, which relates the pH of the buffer solution to the concentration of the acid and its conjugate base:

$$\mathrm{pH} = \mathrm{pK}_a + \log\left(\frac{[\mathrm{A}^-]}{[\mathrm{HA}]}\right)$$

Where:

$[\mathrm{A}^-]$ is the concentration of the conjugate base.

$[\mathrm{HA}]$ is the concentration of the weak acid.

pK_a is the negative logarithm of the acid dissociation constant (K_a).

This equation helps predict the pH of a buffer solution and determine how well it will resist pH changes upon the addition of acids or bases.

4. Applications of Buffer Solutions

Buffer solutions are widely used in various fields, including biology, chemistry, medicine, and environmental science. Here are some key applications:

a. Biological Systems

Cellular Function: Biological systems often operate within a narrow pH range. Buffers maintain this pH, allowing enzymes and biochemical reactions to proceed effectively. For example, the bicarbonate buffer system (H_2CO_3/HCO_3^-) in blood helps maintain physiological pH around 7.4.

Metabolic Processes: During metabolic processes, acids and bases are produced. Buffers like phosphate buffers play critical roles in maintaining intracellular pH, which is essential for cellular function.

b. Industrial Processes

Chemical Manufacturing: Many chemical reactions are sensitive to pH. Buffers are used to maintain the optimal pH for these reactions, ensuring yield and efficiency.

Food Industry: Buffer systems are employed in food preservation and processing. For example, maintaining the pH of food products can prevent spoilage and enhance flavor.

c. Laboratory Applications

pH Control in Experiments: In biochemical and chemical experiments, buffers are crucial for maintaining the desired pH during reactions, which can affect the outcome significantly.

Electrophoresis: Buffers are used in gel electrophoresis to maintain a stable pH, ensuring accurate separation of biomolecules such as DNA and proteins.

d. Environmental Science

Water Quality: Buffers are essential in maintaining the pH of natural water bodies. For instance, bicarbonate buffers in rivers and lakes help neutralize acid rain effects, protecting aquatic ecosystems.

Soil Chemistry: Buffering capacity in soils affects nutrient availability to plants. Farmers often amend soils with lime to adjust pH levels for optimal plant growth.

5. Limitations of Buffers

While buffers are effective in resisting changes in pH, they have limitations:

Capacity: Each buffer has a specific capacity, meaning it can only neutralize a certain amount of acid or base before its pH changes significantly. Once the buffer components are depleted, the solution may undergo drastic pH changes.

Concentration: The effectiveness of a buffer is also dependent on the concentrations of the weak acid and its conjugate base. Higher concentrations generally provide better buffering capacity.

Buffer solutions are critical for maintaining stable pH levels in various chemical and biological systems. Their ability to resist changes in pH upon the addition of acids or bases makes them indispensable in many applications, from biological processes to industrial manufacturing. Understanding how buffers work and their practical uses provides valuable insight into their importance in both laboratory and real-world scenarios. Their role in maintaining the delicate balance of pH is fundamental to life and various technological applications, highlighting their significance in science and industry.

Chapter 9: Thermodynamics and Kinetics

Basics of Thermodynamics: Energy, Heat, and Work

Thermodynamics is a branch of physical chemistry that deals with the study of energy, its transformations, and how it governs the behavior of matter. Understanding the basics of thermodynamics is essential for grasping how chemical reactions and processes occur. The core concepts in this field—**energy**, **heat**, and **work**—are fundamental to understanding how the universe operates at both a macroscopic and molecular level.

1. Energy

Energy is the capacity to do work or transfer heat. It is a fundamental property of matter and exists in various forms such as **kinetic energy** (the energy of motion), **potential energy** (stored energy due to position or configuration), and **thermal energy** (energy related to the temperature of a system). In chemistry, energy changes are crucial because they dictate whether reactions occur spontaneously or require external input.

Kinetic Energy: The energy possessed by moving objects. At the molecular level, it relates to the movement of atoms and molecules.

Potential Energy: This type of energy is stored due to an object's position or arrangement. In chemistry, it is often discussed in terms of bonds between atoms. For example, a chemical bond represents potential energy, which can be released or absorbed during chemical reactions.

In chemical systems, energy is often measured in **joules (J)**, and energy changes during reactions are of prime interest because they tell us how reactions proceed, how much energy is required, and how much energy is released.

2. Heat

Heat is a form of energy transfer between a system and its surroundings due to a temperature difference. Unlike energy, heat refers specifically to the process of energy flow. When two objects at different temperatures come into contact, energy will naturally flow from the hotter object to the cooler one until they reach thermal equilibrium.

In thermodynamics, heat is denoted as **q**, and the sign of q is important:

Positive q means heat is absorbed by the system from the surroundings (endothermic process).

Negative q means heat is released by the system into the surroundings (exothermic process).

Heat can be transferred in three main ways:

Conduction: Transfer of heat through direct contact between molecules.

Convection: Transfer of heat through the movement of fluids (liquids or gases).

Radiation: Transfer of heat through electromagnetic waves, such as sunlight.

Heat is measured in **joules (J)** or **calories (cal)**, where $1\text{cal} = 4.184\text{J}$.

3. Work

Work (W) in a thermodynamic context refers to the process of energy transfer when a force is applied over a distance. In chemistry, the most common type of work encountered is **pressure-volume work** (P-V work), which occurs when gases expand or compress.

The relationship between work and energy in thermodynamics is expressed as:

$$W = -P\Delta V$$

Where:

- P is the external pressure.
- ΔV is the change in volume of the gas.

This equation shows that work is done by a system when its volume changes under pressure. For example:

When a gas expands against external pressure, the system does **work on the surroundings** (work is negative).

When a gas is compressed, **the surroundings do work on the system** (work is positive).

Work, like heat, is measured in joules (J).

4. The First Law of Thermodynamics

The first law of thermodynamics is essentially the **law of energy conservation**, which states that energy cannot be created or destroyed, only transformed from one form to another. In a chemical reaction or physical process, the total energy of a closed system and its surroundings remains constant.

Mathematically, the first law is expressed as:

$$\Delta U = q + W$$

Where:

ΔU is the change in **internal energy** of the system.

q is the heat added to the system.

W is the work done on or by the system.

The internal energy (U) of a system is the total energy contained within the system, including both the kinetic energy of its particles and the potential energy stored in chemical bonds. When energy is transferred as heat or work, the internal energy of the system changes.

5. Internal Energy (U)

Internal energy is the sum of all the microscopic energies within a system. It encompasses the kinetic energy from the motion of molecules, as well as the potential energy from interactions between them, such as intermolecular forces or chemical bonds. Internal energy is a **state function**, meaning its value depends only on the current state of the system (such as temperature, pressure, and composition) and not on how the system arrived at that state.

In practice, we often deal with **changes in internal energy** (ΔU) during chemical reactions. These changes can manifest as heat, work, or both. For example:

In **exothermic reactions**, internal energy decreases as heat is released.

In **endothermic reactions**, internal energy increases as heat is absorbed.

6. The Relationship Between Heat, Work, and Energy in Chemical Reactions

In a chemical reaction, energy changes are observed in the form of heat and work. For example:

In **combustion reactions** (such as burning fuel), energy is primarily released as heat.

In reactions involving gases, work can be done as gases expand or compress.

For an isolated system (where no energy enters or leaves), the total internal energy remains constant. In open or closed systems, energy may be transferred as heat or work, but the total energy change still adheres to the conservation principle.

7. Enthalpy (H)

While internal energy (U) is important, many reactions occur at constant pressure (like most chemical reactions in open systems, such as beakers and flasks). In such cases, chemists often use **enthalpy (H)** to measure energy changes.

Enthalpy is defined as:

$$H = U + PV$$

Where:

H is the enthalpy.

U is the internal energy.

P is pressure.

V is volume.

At constant pressure, the change in enthalpy (ΔH) represents the heat exchanged by the system. This is why chemists frequently refer to changes in enthalpy when discussing heat released or absorbed in reactions.

If ΔH is negative, the reaction is **exothermic** (releases heat).

If ΔH is positive, the reaction is **endothermic** (absorbs heat).

In summary, the basics of thermodynamics—energy, heat, and work—are central to understanding how chemical systems behave. Energy is the driving force behind all physical and chemical processes, while heat and work are mechanisms for transferring energy between systems and their surroundings. The first law of thermodynamics, which highlights energy conservation, provides a foundation for studying how energy is distributed and transformed in chemical reactions. Understanding these principles allows us to predict reaction behavior, measure energy changes, and apply these concepts to a wide range of scientific and industrial processes.

The First and Second Laws of Thermodynamics

Thermodynamics is the branch of chemistry that deals with the relationships between heat, work, and energy in physical and chemical processes. The two fundamental principles governing these processes are the **First Law of Thermodynamics** and the **Second Law of Thermodynamics**. Understanding these laws is essential for grasping how energy is conserved and transformed in chemical reactions, and they have profound implications for both natural phenomena and technological applications.

1. The First Law of Thermodynamics

The First Law of Thermodynamics, often stated as the law of conservation of energy, asserts that energy cannot be created or destroyed; it can only be transformed from one form to another. This fundamental principle can be mathematically expressed as:

$$\Delta U = q + W$$

Where:

ΔU is the change in internal energy of the system.

q is the heat added to the system.

W is the work done on or by the system.

a. **Internal Energy** (U)

Internal energy refers to the total energy contained within a system, encompassing both kinetic and potential energy of its particles. This energy is dependent on several factors, including temperature, volume, and the number of particles present. Internal energy is a state function, meaning it is defined by the state of the system rather than how that state was achieved.

b. Heat (q)

Heat represents the energy transferred between a system and its surroundings due to a temperature difference. When heat is absorbed by the system, q is positive, indicating an endothermic process. Conversely, when the system releases heat to the surroundings, q is negative, indicating an exothermic process.

c. Work (W)

Work is the energy transfer that occurs when a force is applied to an object over a distance. In thermodynamics, the most common type of work considered is pressure-volume work (P-V work), associated with changes in the volume of gases.

When a gas expands against external pressure, the system does work on the surroundings, which is considered negative ($W < 0$).

When a gas is compressed, work is done on the system by the surroundings, which is considered positive ($W > 0$).

The First Law emphasizes that the total energy of an isolated system remains constant. However, energy can flow in and out of the system through heat and work. This law has significant implications in chemical reactions, especially regarding energy changes associated with bond formation and breaking.

Example: In a combustion reaction, the internal energy of the reactants decreases as they form products, releasing heat $(q < 0)$ into the surroundings, resulting in a negative change in internal energy $(\Delta U < 0)$. The energy released can be utilized to do work, such as powering an engine.

2. The Second Law of Thermodynamics

The Second Law of Thermodynamics deals with the direction of energy transformations and introduces the concept of **entropy** (S), which is a measure of the disorder or randomness in a system. The law can be summarized in several ways, but a common interpretation is that:

In any energy transfer or transformation, the total entropy of an isolated system will always increase over time.

This law implies that natural processes tend to move towards a state of greater disorder or randomness.

a. Entropy (S)

Entropy quantifies the dispersal of energy in a system. A system with high entropy has a high degree of disorder, while a system with low entropy is more ordered. The relationship between heat transfer and entropy change can be expressed mathematically as:

$$\Delta S = \frac{q_{rev}}{T}$$

Where:

ΔS is the change in entropy.

q_{rev} is the reversible heat exchanged at a constant temperature.

T is the absolute temperature (in Kelvin).

b. Spontaneous Processes

According to the Second Law, spontaneous processes are characterized by an increase in the total entropy of the universe (system + surroundings). For a process to occur spontaneously, the change in entropy $(\Delta S_{universe})$ must be positive:

$$\Delta S_{universe} = \Delta S_{system} + \Delta S_{surroundings} > 0$$

Example: When ice melts at room temperature, the system (ice and water) absorbs heat from the surroundings, increasing the entropy of the system as solid ice transitions to liquid water. Simultaneously, the surroundings lose heat, but the overall entropy of the universe increases.

c. Applications of the Second Law

The Second Law has significant implications across various fields:

In chemistry, it helps predict whether a reaction will occur spontaneously based on entropy changes.

In engineering, it guides the design of engines and refrigerators by considering efficiency and energy losses.

In biological systems, the Second Law explains processes such as metabolism and the organization of biological molecules.

3. Relationship Between the Laws

The First and Second Laws of Thermodynamics are interconnected:

The First Law ensures energy conservation in any process, while the Second Law addresses the directionality and feasibility of those energy transformations.

Together, these laws provide a comprehensive framework for understanding energy changes and transformations in chemical systems.

The First and Second Laws of Thermodynamics form the cornerstone of thermodynamic principles that govern chemical reactions and physical processes. The First Law emphasizes energy conservation, while the Second Law introduces the concept of entropy and the natural tendency toward disorder. Understanding these laws is essential for studying reaction energetics, predicting the spontaneity of reactions, and applying thermodynamic principles to a wide range of scientific and engineering challenges. These laws not only provide insight into the fundamental nature of energy but also guide practical applications in chemistry, physics, biology, and engineering.

Enthalpy, Entropy, and Gibbs Free Energy

Thermodynamics is a branch of chemistry that explores the relationships between energy, heat, work, and the spontaneity of chemical processes. Three fundamental concepts in thermodynamics are **enthalpy (H)**, **entropy (S)**, and **Gibbs free energy (G)**. These concepts are essential for understanding how energy is transferred and transformed during chemical reactions, and they play a crucial role in predicting reaction behavior and feasibility.

1. Enthalpy (H)

Enthalpy is a thermodynamic property that reflects the total heat content of a system at constant pressure. It combines the internal energy of the system with the energy required to make room for it by displacing its environment, which can be expressed mathematically as:

$$H = U + PV$$

Where:

H is the enthalpy.

U is the internal energy.

P is the pressure.

V is the volume.

a. Change in Enthalpy (ΔH)

The change in enthalpy during a chemical reaction indicates the heat absorbed or released when the reaction occurs at constant pressure. The formula for change in enthalpy is given as:

$$\Delta H = H_{products} - H_{reactants}$$

A positive ΔH (endothermic reaction) signifies that heat is absorbed from the surroundings.

A negative ΔH (exothermic reaction) indicates that heat is released to the surroundings.

b. Applications of Enthalpy

Calorimetry: Enthalpy changes are measured using calorimetry, which quantifies the heat exchanged during chemical reactions.

Thermochemical Equations: These equations represent the enthalpy change associated with reactions, providing valuable information about energy requirements and feasibility.

Enthalpy of Formation: The standard enthalpy change of formation (ΔH_f°) is defined as the change in enthalpy when one mole of a compound is formed from its elements in their standard states. This value is crucial for calculating reaction enthalpies using Hess's Law.

2. Entropy (S)

Entropy is a measure of the disorder or randomness in a system. It quantifies the number of possible configurations that a system can have, reflecting the degree of uncertainty or energy dispersal. The change in entropy for a system can be expressed mathematically as:

$$\Delta S = S_{final} - S_{initial}$$

a. **Understanding Entropy Changes**

Positive ΔS: Indicates an increase in disorder. This often occurs in processes such as melting, vaporization, or the mixing of gases.

Negative ΔS: Indicates a decrease in disorder, which may happen in processes like freezing or the formation of a solid from a solution.

b. Factors Affecting Entropy

Temperature: Entropy increases with temperature. Higher temperatures provide more energy to particles, increasing their movement and disorder.

Phase Changes: Transitioning from solid to liquid or liquid to gas results in higher entropy due to increased molecular freedom.

Molarity and Concentration: Increasing the number of particles in a given volume can lead to higher entropy due to more possible arrangements.

c. Third Law of Thermodynamics

The Third Law of Thermodynamics states that as the temperature approaches absolute zero (0 K), the entropy of a perfect crystal approaches zero. This establishes an absolute reference point for entropy, allowing for more accurate calculations.

3. Gibbs Free Energy (G)

Gibbs free energy is a thermodynamic potential that measures the maximum reversible work obtainable from a thermodynamic system at constant temperature and pressure. It combines enthalpy and entropy to predict the spontaneity of a process:

$$G = H - TS$$

Where:

G is the Gibbs free energy.

T is the absolute temperature (in Kelvin).

S is the entropy.

a. Change in Gibbs Free Energy (ΔG)

The change in Gibbs free energy during a process determines its spontaneity:

$\Delta G < 0$: The process is spontaneous (exergonic).

$\Delta G > 0$: The process is non-spontaneous (endergonic).

$\Delta G = 0$: The system is at equilibrium, and no net change occurs.

This relationship emphasizes that spontaneous processes favor lower energy states and higher disorder.

b. Applications of Gibbs Free Energy

Predicting Reaction Direction: Gibbs free energy provides a criterion for the spontaneity of chemical reactions and processes, enabling chemists to predict whether a reaction will proceed under given conditions.

Phase Equilibria: Gibbs free energy is crucial in understanding phase transitions and equilibrium states.

Biochemical Reactions: In biochemistry, Gibbs free energy is vital for understanding metabolic pathways and enzyme-catalyzed reactions.

4. The Relationship Between Enthalpy, Entropy, and Gibbs Free Energy

The interplay between enthalpy, entropy, and Gibbs free energy is crucial for understanding thermodynamics:

Spontaneous Reactions: Spontaneity depends on both the change in enthalpy and the change in entropy. A reaction can be spontaneous if it releases energy ($\Delta H < 0$) and leads to increased disorder ($\Delta S > 0$).

Temperature Dependence: The temperature plays a significant role in determining spontaneity. As temperature increases, the TS term in the Gibbs free energy equation becomes more significant, potentially leading to non-spontaneity in reactions that are exothermic but have negative entropy changes.

In summary, the concepts of enthalpy, entropy, and Gibbs free energy are foundational in understanding the energy dynamics of chemical reactions and physical processes. Enthalpy reflects heat changes, entropy quantifies disorder, and Gibbs free energy predicts spontaneity and equilibrium. Together, these principles provide a comprehensive framework for analyzing the behavior of chemical systems, guiding both theoretical investigations and practical applications in chemistry, biology, and engineering. Understanding these relationships allows chemists to manipulate conditions to favor desired outcomes in chemical reactions and processes, playing a crucial role in research and industry.

Reaction Rates and Factors Affecting Reaction Rate

Understanding the **rate of chemical reactions** and the factors that influence them is a key part of both thermodynamics and kinetics. Reaction rates tell us how quickly reactants are converted into products in a chemical reaction, which is critical in fields like industrial chemistry, pharmacology, environmental science, and even everyday processes like cooking.

A reaction rate is typically expressed as the change in concentration of a reactant or product over time. It provides insight into how rapidly a reaction proceeds and is determined by both the **kinetic energy** of particles and the number of successful collisions between reactant molecules.

1. Reaction Rates: Definition and Measurement

The **rate of a chemical reaction** refers to how fast reactants are transformed into products. It is mathematically expressed as:

$$\text{Rate} = \frac{\Delta[\text{Product}]}{\Delta t} = -\frac{\Delta[\text{Reactant}]}{\Delta t}$$

Where:

$\Delta[\text{Product}]$ is the change in product concentration.

$\Delta[\text{Reactant}]$ is the change in reactant concentration.

Δt is the change in time.

The negative sign in front of the reactant term signifies that the concentration of reactants decreases over time, while product concentration increases. The rate can be measured in various ways, such as by tracking color change, pressure changes, mass loss, or temperature changes.

2. Factors Affecting Reaction Rates

Several factors can influence the rate of a chemical reaction. These factors alter the frequency and effectiveness of molecular collisions, which directly impacts how quickly a reaction occurs. The major factors include:

a. Concentration of Reactants

The concentration of reactants plays a significant role in determining the reaction rate. According to the **collision theory**, for a reaction to occur, reactant molecules must collide with enough energy. The higher the concentration of reactants, the greater the number of collisions per unit time, which increases the reaction rate. This relationship is particularly true in gases and solutions where the molecules can move freely.

For example:

In a reaction between sodium thiosulfate and hydrochloric acid, increasing the concentration of either reactant will cause the reaction to occur faster due to the higher frequency of collisions.

b. Temperature

Temperature is one of the most crucial factors affecting the rate of a chemical reaction. As temperature increases, the kinetic energy of the molecules also increases, leading to more frequent and more energetic collisions between reactant molecules. This often results in a higher proportion of molecules having sufficient energy to overcome the **activation energy barrier**, leading to a faster reaction rate.

A 10°C increase in temperature typically doubles the rate of many reactions, though the exact increase can vary depending on the specific reaction.

For example, food spoils faster at higher temperatures due to the accelerated rates of chemical reactions involving enzymes and microorganisms.

c. Surface Area of Reactants

In reactions involving solids, the **surface area** of the reactant can significantly influence the reaction rate. When a solid reacts, only the particles on its surface are available for interaction with the other reactants. Increasing the surface area (e.g., by grinding a solid into a fine powder) exposes more particles to potential collisions, thereby increasing the reaction rate.

For example, finely powdered zinc will react faster with hydrochloric acid compared to a large zinc block because more zinc particles are exposed for collision.

d. Catalysts

A **catalyst** is a substance that increases the rate of a chemical reaction without being consumed in the process. Catalysts work by lowering the activation energy required for a reaction, making it easier for reactant molecules to collide with enough energy to form products. This means that even at lower temperatures or concentrations, a reaction can proceed more quickly.

Enzymes are biological catalysts that play a critical role in biochemical reactions in living organisms by speeding up reactions necessary for life processes.

In industry, catalysts are used in the production of chemicals like ammonia in the Haber process, which requires catalysts to be economically viable.

e. Pressure

Pressure primarily affects reactions involving gases. Increasing the pressure of a gas compresses the gas molecules into a smaller volume, leading to a higher concentration of molecules. This results in more frequent collisions and a higher reaction rate. Pressure changes have little to no effect on reactions involving only solids or liquids.

For example, in the industrial synthesis of ammonia (the Haber process), increasing the pressure speeds up the reaction between nitrogen and hydrogen gases.

f. Nature of Reactants

The inherent chemical properties of the reactants themselves also determine the rate of reaction. Some substances naturally react more quickly than others due to factors such as bond strength and molecular complexity. For instance:

Ionic compounds in aqueous solution often react faster than covalent compounds because ions are already dissociated and free to collide.

Simple molecules tend to react faster than complex ones, as complex molecules often require more energy or multiple steps to react.

g. Presence of Inhibitors

In contrast to catalysts, **inhibitors** slow down or prevent chemical reactions. Inhibitors can work by either increasing the activation energy required for a reaction or by blocking the active sites of catalysts. Inhibitors are widely used in industries to control reaction rates or in biological systems to regulate metabolic pathways.

For example, preservatives in food act as inhibitors to slow down the rate of spoilage reactions.

3. The Rate Law and Reaction Order

The **rate law** is an equation that relates the rate of a reaction to the concentration of its reactants. It provides a mathematical expression for how the concentration of each reactant influences the overall reaction rate:

$$\text{Rate} = k[\text{A}]^m[\text{B}]^n$$

Where:

k is the rate constant, which is specific to the reaction and temperature.

$[\text{A}]$ and $[\text{B}]$ are the molar concentrations of reactants A and B.

m and n are the reaction orders for A and B, respectively, and they determine how the concentration of each reactant affects the reaction rate.

The reaction order is determined experimentally and does not necessarily correspond to the stoichiometric coefficients of the balanced chemical equation.

First-order reactions: The rate is directly proportional to the concentration of one reactant.

Second-order reactions: The rate depends on the concentration of two reactants or the square of one reactant.

Zero-order reactions: The rate is independent of the concentration of reactants.

4. Activation Energy and the Arrhenius Equation

Activation energy (E_a) is the minimum energy required for reactants to undergo a chemical reaction. Molecules must possess enough kinetic energy to overcome this barrier in order to react. The relationship between the activation energy, temperature, and reaction rate is described by the **Arrhenius equation**:

$$k = Ae^{-\frac{E_a}{RT}}$$

Where:

k is the rate constant.

A is the frequency factor (related to the number of collisions).

E_a is the activation energy.

R is the gas constant.

T is the temperature in Kelvin.

The Arrhenius equation shows that an increase in temperature lowers the exponential factor, making k larger, which in turn increases the reaction rate.

The rate of a chemical reaction is influenced by a variety of factors, including the concentration of reactants, temperature, surface area, catalysts, and pressure. By understanding these factors, chemists can control the speed of reactions to optimize outcomes in industrial processes, biological systems, and everyday chemical applications. Techniques such as the rate law and the Arrhenius equation offer mathematical insights into how changes in conditions affect reaction rates, making it possible to predict and manipulate chemical behavior efficiently. This knowledge forms the foundation for applications in everything from pharmaceuticals to environmental management.

The Arrhenius Equation and Activation Energy

The **Arrhenius Equation** and **activation energy** are fundamental concepts in chemical kinetics, which explain how temperature influences the rate of a chemical reaction. Together, they provide insights into the molecular collisions and energy thresholds that must be overcome for a reaction to proceed. This understanding is crucial for applications in fields such as catalysis, industrial chemistry, and environmental science, where controlling reaction rates is essential.

1. What is Activation Energy (E□)?

Activation energy (E□)is the minimum amount of energy that reacting molecules must possess to convert from reactants to products during a chemical reaction. It represents the energy barrier that must be overcome for molecules to react. Essentially, not every collision between molecules leads to a reaction; only those collisions where molecules have enough energy to break or form bonds result in a successful reaction. This threshold energy is referred to as the activation energy.

In a potential energy diagram, the activation energy is visualized as the peak of energy between the reactants and products:

Reactants must gain enough energy to reach the top of the energy barrier.

Once they surpass this barrier, the reaction can proceed, resulting in the formation of products.

The higher the activation energy, the fewer the number of reactant molecules that will have sufficient energy to react at a given temperature, and the slower the reaction will be. Conversely, lower activation energy means that more molecules will have enough energy to react, increasing the reaction rate.

For example:

The combustion of wood has a high activation energy, which is why wood does not spontaneously combust at room temperature. However, once you supply sufficient energy (e.g., via a spark or flame), the reaction proceeds rapidly as the activation energy has been overcome.

2. The Arrhenius Equation

The **Arrhenius Equation**, proposed by Swedish chemist Svante Arrhenius in 1889, mathematically describes the relationship between the reaction rate and temperature. It provides a way to calculate the rate constant (k) f a reaction based on the activation energy and the temperature. The equation is given by:

$$k = Ae^{-\frac{E_a}{RT}}$$

Where:

k is the **rate constant** of the reaction.

A is the **frequency factor**, which relates to the number of successful collisions between reactant molecules.

E_a is the **activation energy** of the reaction.

R is the **universal gas constant** (8.314 J/mol·K).

T is the **temperature** in Kelvin.

$e^{-\frac{E_a}{RT}}$ is an exponential term that accounts for the fraction of molecules that possess sufficient energy to overcome the activation energy barrier.

3. Interpretation of the Arrhenius Equation

Each term in the Arrhenius equation has a specific significance in understanding how the rate of a reaction changes with temperature:

a. Rate Constant (k)

The rate constant k is a proportionality factor in the rate law of a chemical reaction. It determines how fast a reaction proceeds under specific conditions. According to the Arrhenius equation, the rate constant increases as temperature rises or as the activation energy decreases, leading to a faster reaction rate.

b. Frequency Factor (A)

The **frequency factor** (A) accounts for the frequency of collisions between reactant molecules and the orientation of molecules during these collisions. It reflects how likely it is for molecules to collide in the correct way for a reaction to occur. This factor depends on the nature of the reactants and the complexity of the reaction. A high value of A means that molecules collide frequently and in the proper orientation, making the reaction more likely to occur.

c. Exponential Term ($e^{-\frac{E_a}{RT}}$)

The exponential term, $\exp\left(-\frac{E_a}{RT}\right)$, represents the fraction of molecules in a system that have enough energy to overcome the activation energy barrier at a given temperature. It reveals the effect of temperature on the reaction rate:

As the temperature (T) increases, the value $\frac{E_a}{RT}$ decreases, which increases the exponential term. This means more molecules have enough energy to react, leading to an increase in the rate constant k and a faster reaction rate.

Conversely, at lower temperatures, the fraction of molecules with sufficient energy decreases, slowing down the reaction.

4. Temperature and Reaction Rates

The Arrhenius equation demonstrates that the rate of a chemical reaction increases with temperature. This is because, at higher temperatures, molecules move faster and have higher kinetic energy. As a result, more molecules have sufficient energy to overcome the activation energy barrier, leading to more successful collisions and a faster reaction.

For many reactions, a **10°C increase in temperature** approximately **doubles the reaction rate**. This general rule is particularly important in both laboratory and industrial settings where controlling temperature can optimize reaction rates.

5. Activation Energy and Catalysis

Catalysts are substances that increase the reaction rate without being consumed in the process. They work by lowering the activation energy E_a of a reaction, making it easier for molecules to overcome the energy barrier and react. Catalysts do this by providing an alternative reaction pathway with a lower activation energy.

Enzymes, which are biological catalysts, are a prime example of how catalysts work in nature. They lower the activation energy of biochemical reactions, allowing essential life processes like metabolism to occur at relatively low temperatures.

In the presence of a catalyst, the value of E_a decreases, and as a result, the exponential term in the Arrhenius equation increases. This means that even at the same temperature, more molecules will have enough energy to react, speeding up the reaction.

For example:

In the decomposition of hydrogen peroxide $(H_2O_2 \rightarrow H_2O + O_2)$, the presence of a catalyst like manganese dioxide (MnO_2) lowers the activation energy and causes the reaction to proceed much more rapidly than without the catalyst.

6. Graphical Representation of the Arrhenius Equation

A useful way to analyze the Arrhenius equation is by taking its natural logarithm:

$$\ln k = \ln A - \frac{E_a}{RT}$$

This equation takes the form of a linear equation $y = mx + b$, where:

$y = \ln k$,

$m = -\frac{E_a}{R}$

$x = \frac{1}{T}$,

$b = \ln A$.

By plotting $\ln k$ against $1/T$ (in Kelvin), you get a **straight line** with a slope of $-E_a/R$. This **Arrhenius plot** is used to determine the activation energy of a reaction from experimental data. The slope of the line provides a way to calculate E_a, and the intercept gives the value of $\ln A$..

7. Applications of the Arrhenius Equation

The Arrhenius equation is fundamental to various practical applications:

a. Industrial Chemistry

In industrial processes, controlling the reaction rate is essential to maximizing product yield and minimizing costs. By adjusting temperature and catalysts, companies can optimize production processes. For example, the

Haber process for ammonia synthesis uses high temperatures and a catalyst to speed up the reaction between nitrogen and hydrogen gases.

b. Pharmaceuticals

In drug manufacturing and storage, the Arrhenius equation is used to predict the **shelf life** of pharmaceuticals. Higher temperatures can increase the rate of chemical degradation, and knowing the activation energy helps estimate how long a drug remains effective.

c. Food Preservation

The Arrhenius equation helps explain why food spoils faster at higher temperatures. By storing food at lower temperatures (e.g., in refrigerators), the activation energy barrier for spoilage reactions is harder to overcome, thus slowing down these reactions and preserving the food for longer periods.

d. Environmental Chemistry

Understanding how temperature affects reaction rates is important in environmental studies, especially in climate modeling and predicting the behavior of atmospheric reactions. For instance, ozone depletion and the formation of smog are temperature-dependent reactions influenced by activation energy.

The Arrhenius equation provides a powerful tool for understanding how temperature and activation energy affect the rates of chemical reactions. By explaining the relationship between temperature and reaction rates, it offers insights into the molecular dynamics of chemical processes. Lowering activation energy through catalysts can greatly enhance reaction rates, leading to practical applications in industries ranging from manufacturing to medicine. Through both theoretical understanding and practical use, the Arrhenius equation continues to be a cornerstone in the study of chemical kinetics.

Chapter 10: Organic Chemistry Basics

Introduction to Organic Chemistry

Organic chemistry is the branch of chemistry that focuses on the study of carbon-containing compounds and their structures, properties, reactions, and synthesis. It plays a critical role in the modern world, forming the backbone of industries such as pharmaceuticals, biotechnology, petrochemicals, food, and materials science. Organic compounds are ubiquitous, found in everything from the food we eat to the clothes we wear, and even the air we breathe.

1. What is Organic Chemistry?

Organic chemistry primarily studies compounds composed of carbon atoms, often in combination with hydrogen, oxygen, nitrogen, sulfur, phosphorus, and halogens (such as chlorine, bromine, iodine, and fluorine). This vast field encompasses both naturally occurring substances (like proteins, fats, and carbohydrates) and synthetic compounds (such as plastics, pharmaceuticals, and dyes).

The reason carbon is central to organic chemistry lies in its unique bonding properties:

Carbon can form **four covalent bonds** with other atoms, creating complex and stable molecules.

It can bond to itself, forming long chains, rings, and branched structures, which is rare among elements.

These diverse structures lead to an enormous variety of organic molecules with different functional groups and properties.

2. Historical Background of Organic Chemistry

The field of organic chemistry has its roots in the early 19th century when scientists were trying to understand substances derived from living organisms. For a long time, it was believed that organic compounds could only be produced by living organisms, a concept known as **vitalism**. This theory was debunked in 1828 when German chemist Friedrich Wöhler synthesized **urea** (an organic compound) from an inorganic compound, ammonium cyanate, marking the beginning of modern organic chemistry.

This discovery led to the realization that organic compounds could be synthesized artificially in the laboratory, expanding the scope of chemistry from naturally occurring substances to man-made materials.

3. Key Characteristics of Organic Compounds

Organic compounds are incredibly diverse, but they share several common features that make them distinct from inorganic compounds:

a. Carbon Framework

The key feature of organic molecules is their **carbon skeleton**. Carbon atoms can bond to form chains (linear, branched, or cyclic) that serve as the backbone for the molecule. This ability allows for immense structural diversity, from small molecules like methane (CH_4) to complex macromolecules like proteins and DNA.

b. Hydrocarbons

Many organic compounds are **hydrocarbons**, meaning they are composed solely of carbon and hydrogen atoms. Hydrocarbons can be divided into:

Alkanes (saturated hydrocarbons with single bonds),

Alkenes (unsaturated hydrocarbons with double bonds),

Alkynes (unsaturated hydrocarbons with triple bonds), and

Aromatic hydrocarbons (compounds containing benzene rings or similar structures).

c. Functional Groups

A hallmark of organic chemistry is the presence of **functional groups** — specific atoms or groups of atoms within a molecule that dictate its chemical behavior. Some of the most common functional groups include:

Hydroxyl group (-OH): Found in alcohols,

Carbonyl group (C=O): Found in aldehydes and ketones,

Carboxyl group (-COOH): Found in carboxylic acids,

Amino group ($-NH_2$): Found in amines and amino acids.

These functional groups determine the reactivity and interactions of organic molecules, making them essential for understanding the vast range of organic reactions.

d. Covalent Bonding

In organic compounds, atoms are typically linked by **covalent bonds**, where electrons are shared between atoms. The strength and versatility of these bonds, particularly those involving carbon, allow for the stability and flexibility of organic molecules.

4. The Importance of Organic Chemistry

Organic chemistry is fundamental to many aspects of life and industry. Its importance can be understood by looking at a few key areas:

a. Biological Systems

Organic compounds are the building blocks of life. The molecules that make up living organisms — including **proteins, nucleic acids (DNA and RNA), carbohydrates, lipids**, and **enzymes** — are organic. Understanding the structure and reactions of these biomolecules is crucial for studying biological processes, genetics, and physiology.

For example:

Proteins are made of amino acids, which contain both amine and carboxyl groups.

DNA is composed of a sugar-phosphate backbone and nitrogenous bases, all of which are organic in nature.

Enzymes, which catalyze biochemical reactions, are organic molecules essential for metabolism.

b. Pharmaceuticals and Medicine

The development of drugs relies heavily on organic chemistry. Most pharmaceuticals are organic compounds designed to interact with biological targets (like enzymes or receptors) to treat or prevent diseases. Organic chemists play a key role in drug design, synthesis, and testing, contributing to advancements in medicine.

For example, **penicillin**, one of the first antibiotics, is an organic compound that revolutionized healthcare. More recent drugs, like **statins** (used to lower cholesterol) and **antiretrovirals** (used to treat HIV/AIDS), are also products of organic chemistry.

c. Agriculture

Organic chemistry has transformed agriculture through the development of **pesticides, herbicides, fertilizers, and fungicides**. These compounds help improve crop yields and protect plants from pests and diseases, ensuring food security for a growing global population.

d. Materials and Polymers

The field of organic chemistry is responsible for the creation of **polymers** — large molecules composed of repeating units of smaller molecules called monomers. Polymers include everyday materials like **plastics, rubbers, and fibers** used in packaging, clothing, electronics, and construction.

For example:

Polyethylene is used in plastic bags and bottles.

Nylon is a synthetic polymer used in textiles.

Kevlar is an organic polymer known for its high strength and used in bulletproof vests.

e. Energy

Many energy sources are derived from organic compounds, particularly **fossil fuels** like coal, oil, and natural gas. The combustion of hydrocarbons in these fuels produces energy, which powers industries and transportation. Organic chemistry is also key to developing **biofuels** and exploring more sustainable sources of energy, such as **solar cells** and **batteries**.

5. The Scope of Organic Chemistry

The scope of organic chemistry is vast, covering everything from simple hydrocarbons to complex biomolecules and synthetic compounds. Some important areas of focus within organic chemistry include:

a. Synthesis

Organic synthesis is the process of creating organic compounds from simpler starting materials. It is one of the central tasks of organic chemists and is crucial for creating new pharmaceuticals, materials, and chemicals. Synthesis involves combining various reactions, such as **addition, substitution, elimination, and rearrangement reactions**, to build complex molecules.

b. Reaction Mechanisms

Understanding **reaction mechanisms** — the step-by-step process by which a chemical reaction occurs — is a key aspect of organic chemistry. Reaction mechanisms explain how and why bonds are formed or broken during a chemical reaction, shedding light on the molecular behavior that leads to the observed products.

c. Spectroscopy and Structural Determination

Organic chemists use **spectroscopy** to analyze and determine the structure of organic compounds. Techniques like **Nuclear Magnetic Resonance (NMR), Infrared (IR) spectroscopy**, and **Mass Spectrometry (MS)** help identify the functional groups, bonding patterns, and molecular weights of organic molecules.

d. Environmental Organic Chemistry

Organic chemistry is also critical in understanding and addressing environmental issues. Organic pollutants, such as **pesticides, herbicides**, and **industrial chemicals**, can have harmful effects on ecosystems and human health. Organic chemists study these compounds to develop ways to degrade them or mitigate their impact.

Organic chemistry is a central and dynamic field that underpins much of modern science and technology. From the molecular mechanisms of life to the design of new drugs and materials, organic chemistry is vital for understanding and advancing a wide range of industries. Its study opens the door to innovations in health, energy, agriculture, and environmental sustainability. Whether you're looking to explore the world of molecules or create novel compounds, organic chemistry offers endless opportunities for discovery and application.

Hydrocarbons: Alkanes, Alkenes, Alkynes, and Aromatic Compounds

Hydrocarbons are organic compounds consisting entirely of carbon (C) and hydrogen (H) atoms. They serve as the fundamental building blocks of organic chemistry, forming the core structures from which more complex organic compounds are derived. Hydrocarbons are classified into various categories based on the type of bonds present between the carbon atoms. The main categories include alkanes, alkenes, alkynes, and aromatic compounds. Each of these classes has distinct characteristics, structures, and properties that play a crucial role in both natural processes and industrial applications.

1. Alkanes

Alkanes, also known as paraffins, are saturated hydrocarbons characterized by single carbon-carbon (C–C) bonds. C_nH_{2n+2}, where n is the number of carbon atoms.

a. Structure and Properties

Structure: Alkanes can exist in straight-chain (linear) or branched forms. For example, butane can be represented as either a straight chain or as isobutane (a branched version).

Properties: Alkanes are typically nonpolar, making them insoluble in water but soluble in nonpolar solvents. They have relatively low reactivity compared to other hydrocarbons due to the strength of the C–C and C–H bonds. As a result, they primarily undergo reactions such as combustion and substitution.

b. Examples of Alkanes

Methane (CH_4): The simplest alkane, a major component of natural gas, and an important energy source.

Ethane (C_2H_6): Used in the production of ethylene and as a fuel.

Propane (C_3H_8): Commonly used as a fuel for heating and cooking.

Butane (C_4H_{10}): Often used in lighters and as a fuel in portable stoves.

2. Alkenes

Alkenes, or **olefins**, are unsaturated hydrocarbons that contain at least one carbon-carbon double bond (C=C). Their general formula is C_nH_{2n}.

a. Structure and Properties

Structure: The presence of a double bond introduces a planar geometry around the carbon atoms involved, leading to restricted rotation and the possibility of cis-trans (geometric) isomerism.

Properties: Alkenes are more reactive than alkanes due to the presence of the double bond. They readily undergo addition reactions, where atoms or groups are added across the double bond.

b. Examples of Alkenes

Ethylene (C_2H_4): Widely used as a precursor for the production of plastics, antifreeze, and synthetic fibers.

Propylene (C_3H_6): A key building block in the manufacture of polypropylene, a common plastic.

Butene (C_4H_8): Used in the production of gasoline and as a chemical intermediate.

3. Alkynes

Alkynes are unsaturated hydrocarbons that contain at least one carbon-carbon triple bond (C≡C). Their general formula is C_nH_{2n-2}.

a. Structure and Properties

Structure: Alkynes have a linear arrangement of atoms around the triple bond, resulting in a more elongated structure. The triple bond is stronger and shorter than a double bond, contributing to the unique properties of alkynes.

Properties: Like alkenes, alkynes are reactive and can undergo addition reactions, but they can also participate in elimination reactions. They are also used in various industrial applications.

b. Examples of Alkynes

Acetylene (C_2H_2): The simplest alkyne, commonly used as a fuel in welding and cutting metals.

Propyne (C_3H_4): Used in organic synthesis and as a fuel gas.

Butyne (C_4H_6): Utilized in chemical synthesis and as an intermediate in the production of other chemicals.

4. Aromatic Compounds

Aromatic compounds are a distinct class of hydrocarbons characterized by the presence of one or more **benzene rings** in their structure. Benzene (C_6H_6) is the simplest aromatic compound and serves as the prototype for this class.

a. Structure and Properties

Structure: Aromatic compounds are typically cyclic and planar, with alternating double bonds between carbon atoms. This delocalization of electrons gives rise to the concept of resonance, where the electrons are shared across the entire ring structure.

Properties: Aromatic compounds are generally more stable than their non-aromatic counterparts due to resonance. They tend to be less reactive than alkenes and alkynes but can undergo substitution reactions (where one hydrogen atom is replaced by another atom or group) rather than addition reactions.

b. Examples of Aromatic Compounds

Benzene (C_6H_6): A fundamental building block in organic chemistry, used in the production of various chemicals, including plastics, dyes, and pharmaceuticals.

Toluene (C_7H_8): Commonly used as a solvent and in the production of chemicals like benzene derivatives.

Xylene (C_8H_{10}): Often used as a solvent in laboratories and as an intermediate in the production of various chemicals.

5. Summary of Hydrocarbon Classes

Alkanes: Saturated, single bonds (C–C), general formula C_nH_{2n+2}.

Alkenes: Unsaturated, at least one double bond (C=C), general formula C_nH_{2n}.

Alkynes: Unsaturated, at least one triple bond (C≡C), general formula C_nH_{2n-2}.

Aromatic Compounds: Contain one or more benzene rings, exhibit resonance stability.

Hydrocarbons form the backbone of organic chemistry and are critical for understanding the structure, behavior, and reactions of organic compounds. From simple alkanes to complex aromatic compounds, each class of hydrocarbons has unique properties and applications that significantly impact industries such as pharmaceuticals, agriculture, and materials science. The study of hydrocarbons not only provides insight into fundamental chemical principles but also lays the groundwork for more complex organic chemistry concepts and applications. Understanding these hydrocarbons is essential for anyone pursuing a career in chemistry or related fields.

Functional Groups: Alcohols, Ethers, Aldehydes, Ketones, and Carboxylic Acids

Functional groups are specific groups of atoms within molecules that are responsible for the characteristic chemical reactions of those molecules. Understanding functional groups is crucial for grasping organic chemistry, as they determine the properties, reactivity, and behavior of organic compounds. This chapter will explore five important functional groups: **alcohols, ethers, aldehydes, ketones, and carboxylic acids**. Each of these functional groups plays a significant role in biological processes, industrial applications, and the synthesis of pharmaceuticals.

1. Alcohols

Alcohols are organic compounds that contain one or more hydroxyl (-OH) functional groups attached to a carbon atom. The general formula for alcohols is $C_nH_{2n+1}OH$.

a. Structure and Properties

Structure: The presence of the hydroxyl group (-OH) classifies a compound as an alcohol. The carbon atom to which the hydroxyl group is attached can be primary, secondary, or tertiary, depending on the number of carbon atoms bonded to it.

Properties: Alcohols are polar due to the hydroxyl group, which makes them soluble in water. They have higher boiling points than hydrocarbons of similar molecular weight due to hydrogen bonding. Alcohols can act as weak acids, donating protons in solution.

b. Examples of Alcohols

Methanol (CH_3OH): Commonly known as wood alcohol, methanol is used as an industrial solvent, antifreeze, and fuel.

Ethanol (C_2H_5OH): The type of alcohol found in alcoholic beverages, ethanol is also used as a solvent and in the manufacture of personal care products.

Isopropanol (C_3H_8O): Also known as isopropyl alcohol, it is commonly used as a disinfectant and solvent.

2. Ethers

Ethers are organic compounds in which an oxygen atom is connected to two alkyl or aryl groups. The general structure of an ether is R-O-R', where R and R' represent hydrocarbon chains.

a. Structure and Properties

Structure: The oxygen atom in an ether is bonded to two carbon atoms, which can be the same (symmetrical ethers) or different (asymmetrical ethers).

Properties: Ethers are generally nonpolar and have lower boiling points compared to alcohols because they cannot form hydrogen bonds with themselves. However, ethers can participate in hydrogen bonding with water, making some ethers moderately soluble.

b. Examples of Ethers

Dimethyl Ether (C_2H_6O): A simple ether that can be used as a fuel and in the synthesis of other chemicals.

Ethyl Ether (Diethyl Ether, $C_4H_{10}O$): Commonly used as a solvent and historically used as an anesthetic.

3. Aldehydes

Aldehydes are organic compounds containing a carbonyl group (C=O) bonded to at least one hydrogen atom. The general formula for aldehydes is $RCHO.$, where R is a hydrocarbon chain.

a. Structure and Properties

Structure: The carbonyl group in aldehydes is always located at the end of the carbon chain, making them distinct from ketones.

Properties: Aldehydes are polar due to the carbonyl group, which makes them soluble in water, particularly if the R group is small. They have relatively higher boiling points than hydrocarbons but lower than alcohols. Aldehydes are generally reactive and can undergo oxidation to form carboxylic acids.

b. Examples of Aldehydes

Formaldehyde (HCHO): A colorless gas with a pungent odor, used as a preservative and in the manufacture of resins.

Acetaldehyde (C_2H_4O): Used in the production of acetic acid and as a flavoring agent.

4. Ketones

Ketones are organic compounds that contain a carbonyl group (C=O) bonded to two carbon atoms. The general formula for ketones is $R_2C = O.$

a. Structure and Properties

Structure: Unlike aldehydes, ketones have the carbonyl group located within the carbon chain, making them more stable and less reactive.

Properties: Ketones are polar and have higher boiling points than alkanes but lower than alcohols. They are also soluble in water and can participate in hydrogen bonding.

b. Examples of Ketones

Acetone (C_3H_6O): The simplest ketone, widely used as a solvent in nail polish remover and in laboratories.

Butanone (C_4H_8O): Also known as methyl ethyl ketone (MEK), it is used as a solvent in various industrial processes.

5. Carboxylic Acids

Carboxylic acids are organic compounds that contain a carboxyl group (-COOH), which is a combination of a carbonyl group and a hydroxyl group. The general formula for carboxylic acids is $RCOOH.$

a. Structure and Properties

Structure: The carboxyl group gives these acids their acidic properties. The presence of the hydroxyl group means that carboxylic acids can donate protons, making them acidic.

Properties: Carboxylic acids are polar and have higher boiling points than alcohols and aldehydes due to the ability to form hydrogen bonds. They are also soluble in water, especially if the carbon chain is short.

b. Examples of Carboxylic Acids

Formic Acid (HCOOH): The simplest carboxylic acid, found in ant venom and used in leather production.

Acetic Acid (CH_3COOH): Commonly known as vinegar, acetic acid is widely used in food preservation and chemical synthesis.

Benzoic Acid ($C_7H_6O_2$): Used as a preservative and in the manufacture of various chemicals.

6. Summary of Functional Groups

Alcohols: Contain hydroxyl (-OH) groups, generally polar and soluble in water.

Ethers: Have an oxygen atom connected to two carbon chains, generally nonpolar with lower boiling points.

Aldehydes: Contain a carbonyl group (C=O) at the end of the carbon chain, polar and reactive.

Ketones: Contain a carbonyl group within the carbon chain, more stable and less reactive than aldehydes.

Carboxylic Acids: Contain a carboxyl group (-COOH), polar and acidic due to the ability to donate protons.

Understanding functional groups is essential in organic chemistry as they dictate the reactivity, solubility, and properties of organic molecules. Alcohols, ethers, aldehydes, ketones, and carboxylic acids represent key functional groups that are prevalent in a wide variety of chemical reactions and biological processes. This knowledge forms the foundation for further exploration of organic compounds and their applications in pharmaceuticals, agriculture, and material science. Mastery of functional groups not only enhances one's understanding of organic chemistry but also provides the tools necessary for innovation in chemical synthesis and application.

Basic Organic Reactions

Organic chemistry is a vast and intricate field that focuses on the study of carbon-containing compounds and their reactions. At its core, organic chemistry involves understanding how these compounds interact, transform, and yield new substances through chemical reactions. This chapter delves into the basic types of organic reactions, their mechanisms, and the significance of these reactions in both laboratory settings and real-world applications.

1. Introduction to Organic Reactions

Organic reactions can be broadly classified into several categories based on their mechanisms and the types of chemical changes that occur. Understanding these categories is crucial for predicting the behavior of organic compounds and for designing new synthetic pathways in chemistry.

2. Types of Basic Organic Reactions

The fundamental types of organic reactions include:

a. Addition Reactions

Addition reactions occur when two reactants combine to form a single product. This type of reaction is common with unsaturated compounds, such as alkenes and alkynes, which contain carbon-carbon double or triple bonds.

Mechanism: In an addition reaction, the π bond of the alkene or alkyne breaks, allowing new atoms or groups to be added to the carbon atoms.

Example:

Hydrogenation: The addition of hydrogen (H_2) across the double bond of an alkene, converting it into an alkane.

Hydration: The addition of water (H_2O) to an alkene in the presence of an acid catalyst, forming an alcohol.

b. Substitution Reactions

Substitution reactions involve the replacement of one atom or group in a molecule with another atom or group. These reactions are common in saturated compounds like alkanes and aromatic compounds.

Mechanism: A leaving group departs from the molecule, and a new group takes its place. Substitution reactions can be classified as either nucleophilic or electrophilic.

Example:

Nucleophilic Substitution: In halogenated hydrocarbons, a halogen atom (like bromine) can be replaced by a nucleophile (like OH^-), resulting in the formation of alcohol.

Electrophilic Aromatic Substitution: In aromatic compounds, a hydrogen atom can be replaced by an electrophile, such as in the nitration of benzene to form nitrobenzene.

c. Elimination Reactions

Elimination reactions involve the removal of a small molecule from a larger one, often resulting in the formation of a double bond or triple bond. These reactions are the reverse of addition reactions.

Mechanism: In an elimination reaction, the departing groups must be positioned correctly for the reaction to occur, often leading to the formation of a π bond.

Example:

Dehydration: The elimination of water from alcohols to form alkenes.

Dehydrohalogenation: The removal of hydrogen halides (like HCl) from alkyl halides to yield alkenes.

d. Rearrangement Reactions

Rearrangement reactions involve the structural rearrangement of a molecule, leading to the formation of an isomer. These reactions often occur through the migration of atoms or groups within the molecule.

Mechanism: A bond breaks and a new bond forms, resulting in a change of the molecule's connectivity.

Example:

Hydride Shift: In the rearrangement of carbocations, a hydride ion can migrate from one carbon to an adjacent carbon, resulting in a more stable carbocation.

3. Factors Affecting Organic Reactions

Several factors influence the course of organic reactions, including:

a. Structure of the Reactants

The nature and arrangement of atoms in the reactants play a crucial role in determining reaction pathways. For example, steric hindrance can prevent certain reactions from occurring.

b. Reaction Conditions

Temperature, pressure, and the presence of catalysts can significantly affect reaction rates and mechanisms. For example, increasing temperature usually increases the kinetic energy of the molecules, leading to faster reaction rates.

c. Solvent Effects

The choice of solvent can influence the reaction mechanism and the stability of intermediates. Polar solvents may stabilize charged intermediates, while nonpolar solvents may favor different pathways.

4. Examples of Organic Reactions

a. Hydrogenation of Alkenes

Reaction:

$$RCH = CHR' + H_2 \xrightarrow{Pd/C} RCH_2CH_2R'$$

In this reaction, hydrogen is added across the double bond of an alkene in the presence of a catalyst like palladium on carbon, resulting in the corresponding alkane.

b. Nitration of Benzene

Reaction:

$$C_6H_6 + HNO_3 \xrightarrow{H_2SO_4} C_6H_5NO_2 + H_2O$$

- In this electrophilic aromatic substitution reaction, benzene reacts with nitric acid in the presence of sulfuric acid to form nitrobenzene.

5. Importance of Basic Organic Reactions

Understanding basic organic reactions is essential for several reasons:

a. Synthesis of Organic Compounds

Organic reactions provide the foundational techniques needed to synthesize a wide range of organic compounds, from simple molecules to complex pharmaceuticals.

b. Understanding Biological Processes

Many biological processes are driven by organic reactions. For instance, enzymatic reactions in metabolism involve addition, substitution, and elimination processes that are vital for sustaining life.

c. Environmental Chemistry

Organic reactions are fundamental to understanding environmental processes, such as the degradation of pollutants and the formation of naturally occurring compounds in ecosystems.

Basic organic reactions form the building blocks of organic chemistry. By understanding the types of reactions, their mechanisms, and the factors that influence them, students can develop a comprehensive understanding of how organic compounds behave and interact. This knowledge not only serves as a foundation for more advanced studies in organic chemistry but also equips students with the tools necessary to tackle real-world problems in chemical synthesis, pharmaceuticals, and environmental science. Mastery of these fundamental concepts is essential for any aspiring chemist or scientist interested in the intricate world of organic chemistry.

Isomerism: Structural and Stereoisomerism

Isomerism is a fundamental concept in organic chemistry that refers to the phenomenon where two or more compounds have the same molecular formula but different structures or arrangements of atoms. This chapter will explore the two main types of isomerism: structural isomerism and stereoisomerism. Understanding isomerism is crucial because even slight differences in the arrangement of atoms can lead to significant variations in physical and chemical properties.

1. Introduction to Isomerism

Isomers are compounds that share the same molecular formula (i.e., the same number of atoms of each element) but differ in structure or spatial arrangement. This can result in diverse chemical behaviors, reactivities, and biological activities, making isomerism a key concept in organic chemistry.

2. Types of Isomerism

Isomerism is generally categorized into two main types: structural isomerism and stereoisomerism.

A. Structural Isomerism

Structural isomerism arises when isomers differ in the connectivity of their atoms. This type of isomerism can be further classified into several subcategories:

1. Chain Isomerism

Chain isomerism occurs when the isomers differ in the arrangement of the carbon skeleton. For instance, they may have branched or straight-chain structures.

Example: Butane (C_4H_{10}) can exist as:

n-butane: a straight-chain alkane (CH_3-CH_2-CH_2-CH_3)

isobutane: a branched-chain alkane ($(CH_3)_2CH$-)

2. Position Isomerism

Position isomerism occurs when the position of a functional group, double bond, or substituent changes on the carbon chain.

Example: Butanol can exist as:

1-butanol: CH_3-CH_2-CH_2-CH_2OH (hydroxyl group on the first carbon)

2-butanol: CH_3-CH_2-CHOH-CH_3 (hydroxyl group on the second carbon)

3. Functional Group Isomerism

Functional group isomerism occurs when compounds with the same molecular formula belong to different functional groups.

Example: C_4H_8O can be represented as:

Butanal: an aldehyde (CH_3-CH_2-CH_2-CHO)

Butan-2-ol: an alcohol (CH_3-CH_2-CHOH-CH_3)

4. Tautomeric Isomerism

Tautomeric isomerism is a special type of functional group isomerism that involves the migration of a hydrogen atom and a shift of a double bond.

Example: The keto-enol tautomerism of acetone (CH_3COCH_3) can exist as:

Keto form: CH_3COCH_3

Enol form: CH_2=CHOH

B. Stereoisomerism

Stereoisomerism occurs when isomers have the same molecular formula and connectivity of atoms but differ in the spatial arrangement of their atoms. This can be further divided into two main categories: geometric (cis-trans) isomerism and optical isomerism.

1. Geometric Isomerism (Cis-Trans Isomerism)

Geometric isomers occur due to the restricted rotation around a double bond or a ring structure, leading to different spatial arrangements of substituents.

Cis isomer: Similar groups are on the same side of the double bond or ring.

Trans isomer: Similar groups are on opposite sides.

Example:

Cis-2-butene: CH_3-CH=CH-CH_3 (methyl groups on the same side)

Trans-2-butene: CH_3-CH=CH-CH_3 (methyl groups on opposite sides)

2. Optical Isomerism

Optical isomerism arises from the presence of chiral centers—atoms, usually carbon, that are bonded to four different substituents. This results in non-superimposable mirror images called enantiomers.

Enantiomers: Two stereoisomers that are mirror images of each other but cannot be aligned perfectly (like left and right hands).

Chirality: The property of being chiral, often due to the presence of a chiral center.

Example:

Lactic acid ($C_3H_6O_3$) has two enantiomers:

D-lactic acid: (S)-lactic acid

L-lactic acid: (R)-lactic acid

3. Importance of Isomerism in Chemistry

Isomerism plays a critical role in organic chemistry and has significant implications in various fields:

A. Pharmaceuticals

Many drugs are chiral, meaning their effectiveness can vary significantly between enantiomers. For instance, one enantiomer of a drug may have therapeutic effects, while the other may be inactive or even harmful.

Example: The drug thalidomide had one enantiomer that was effective against morning sickness, while the other caused severe birth defects.

B. Biochemistry

Isomerism is vital in biochemistry as many biological molecules, such as amino acids and sugars, exhibit isomerism. The specific three-dimensional arrangement of these molecules is crucial for their function.

C. Material Science

In material science, different isomers can lead to variations in physical properties, such as melting points, solubility, and stability, which are critical for the development of new materials.

Isomerism is a central concept in organic chemistry that enhances our understanding of the diversity of organic compounds. By recognizing the various types of isomerism—structural and stereoisomerism—chemists can predict the properties, reactivities, and behaviors of compounds. This knowledge not only aids in the synthesis and development of new organic molecules but also has far-reaching implications in pharmaceuticals, biochemistry, and materials science. Understanding isomerism is essential for any chemist seeking to navigate the intricate landscape of organic chemistry and its applications.

Chapter 11: Inorganic Chemistry Essentials

Introduction to Inorganic Chemistry

Inorganic chemistry is a branch of chemistry that deals with the properties and behavior of inorganic compounds, which include minerals, metals, and all compounds that are not classified as organic. While organic chemistry primarily focuses on carbon-based compounds, inorganic chemistry covers a broader range of substances and plays a crucial role in various fields, including materials science, catalysis, environmental science, and biological processes.

1. Definition and Scope of Inorganic Chemistry

Inorganic chemistry encompasses a wide variety of compounds that can be divided into several categories, including:

Metallic Compounds: These are composed of metals and often display metallic properties such as conductivity and malleability. Examples include transition metals and their complexes.

Non-Metallic Compounds: These include compounds formed by non-metals, such as halides, oxides, and sulfides. Non-metallic elements often exhibit diverse properties and behaviors.

Coordination Compounds: These consist of a central metal atom bonded to surrounding ligands (molecules or ions that can donate electron pairs). Coordination compounds are significant in various biological systems and industrial applications.

Organometallic Compounds: These compounds contain metal-carbon bonds, bridging the gap between organic and inorganic chemistry. They are critical in catalysis and materials science.

2. Historical Background

Inorganic chemistry has a rich historical background, dating back to ancient times when humans began to extract and manipulate metals and minerals. Key developments in the field include:

Alchemical Traditions: The early study of inorganic substances was rooted in alchemy, where practitioners sought to transform base metals into gold and discover the philosopher's stone. While these pursuits were largely mystical, they laid the groundwork for modern chemistry.

Discovery of Elements: The identification of various elements in the 18th and 19th centuries, such as oxygen, hydrogen, and sulfur, significantly advanced the field. Dmitri Mendeleev's periodic table was a pivotal moment, organizing elements based on their atomic mass and properties.

Development of Coordination Chemistry: The 19th century also saw significant advances in coordination chemistry, with scientists like Alfred Werner pioneering the understanding of complex formation and coordination compounds.

3. Key Concepts in Inorganic Chemistry

a. The Periodic Table

The periodic table is a fundamental tool in inorganic chemistry, organizing elements based on their atomic number, electron configuration, and recurring chemical properties. Understanding periodic trends—such as electronegativity, atomic radius, ionization energy, and metallic character—is crucial for predicting the behavior of inorganic compounds.

b. Chemical Bonding

Inorganic chemistry involves various types of chemical bonding, including:

Ionic Bonding: Involves the transfer of electrons from one atom to another, resulting in the formation of charged ions. Ionic compounds typically have high melting and boiling points.

Covalent Bonding: Involves the sharing of electrons between atoms, forming molecules with specific geometries. This type of bonding is common in non-metallic compounds.

Metallic Bonding: Occurs in metals, where electrons are delocalized, allowing for conductivity and malleability.

Understanding these bonding types helps explain the properties and behaviors of inorganic substances.

c. Acids and Bases

Inorganic chemistry includes various theories of acids and bases, such as:

Arrhenius Theory: Defines acids as substances that produce hydrogen ions (H^+) in aqueous solutions, while bases produce hydroxide ions (OH^-).

Brønsted-Lowry Theory: Expands on this by defining acids as proton donors and bases as proton acceptors, applicable to non-aqueous systems.

Lewis Theory: Defines acids as electron pair acceptors and bases as electron pair donors, emphasizing the role of electron transfer in chemical reactions.

4. Applications of Inorganic Chemistry

Inorganic chemistry has numerous practical applications across various fields, including:

a. Materials Science

Inorganic compounds, such as ceramics, metals, and semiconductors, play a crucial role in the development of new materials. Inorganic chemists design materials with specific properties for applications in electronics, construction, and manufacturing.

b. Catalysis

Many inorganic compounds serve as catalysts in chemical reactions, speeding up reactions without being consumed. Transition metals and their complexes are particularly valuable in catalyzing industrial processes, such as hydrogenation and polymerization.

c. Environmental Chemistry

Inorganic chemistry is vital in understanding environmental processes, including pollution control, water treatment, and the cycling of nutrients in ecosystems. Inorganic chemists study the behavior of metals and non-metals in the environment, helping to mitigate environmental impact.

d. Biological Inorganic Chemistry

Inorganic compounds are essential to biological systems. Elements like iron, zinc, copper, and magnesium play critical roles in biological processes, including oxygen transport, enzyme activity, and photosynthesis. Understanding these roles is essential for fields like biochemistry and medicine.

5. Current Trends and Future Directions

Inorganic chemistry continues to evolve, with current research focusing on:

Nanotechnology: The development and application of inorganic nanoparticles in fields like drug delivery, imaging, and catalysis.

Green Chemistry: Designing inorganic reactions and processes that minimize waste and reduce environmental impact, such as using non-toxic solvents and renewable resources.

Sustainable Materials: Developing inorganic materials that are environmentally friendly and sustainable, including biodegradable and recyclable substances.

Inorganic chemistry is a fundamental branch of chemistry that offers insight into the behavior and properties of a wide range of compounds. Its applications span numerous fields, from materials science to environmental chemistry, making it a vital area of study. As new technologies and methods emerge, the importance of inorganic chemistry will only continue to grow, shaping our understanding of the natural world and driving innovations for the future. Understanding the principles and concepts of inorganic chemistry is essential for anyone pursuing a career in chemistry, environmental science, or related fields.

Coordination Chemistry and Complexes

Coordination chemistry is a significant branch of inorganic chemistry that focuses on the structures, properties, and reactions of coordination complexes. These complexes consist of a central metal atom or ion surrounded by molecules or anions known as ligands. Coordination chemistry plays a crucial role in various fields, including catalysis, materials science, bioinorganic chemistry, and the development of pharmaceuticals. Understanding the principles of coordination chemistry provides insights into the behavior of metal ions in different environments, both in nature and in synthetic applications.

1. Definition of Coordination Compounds

Coordination compounds, also known as coordination complexes, consist of a central metal atom or ion bonded to one or more ligands. Ligands are molecules or ions that can donate electron pairs to the metal, forming coordinate covalent bonds. These complexes exhibit unique chemical properties that differ from those of the individual metal ions and ligands.

a. Central Metal Atom or Ion

The central metal atom is typically a transition metal due to their ability to form multiple oxidation states and complex geometries. Common examples include iron (Fe), copper (Cu), nickel (Ni), and cobalt (Co).

b. Ligands

Ligands are classified based on the number of binding sites they possess:

Monodentate Ligands: Ligands that attach to the metal through a single atom. Examples include water (H_2O), ammonia (NH_3), and chloride ions (Cl^-).

Bidentate Ligands: Ligands that can bond through two atoms. An example is ethylenediamine (en), which binds through both nitrogen atoms.

Polydentate Ligands: Ligands with multiple binding sites. A classic example is ethylenediaminetetraacetic acid (EDTA), which can coordinate through four carboxylate groups and two amine groups.

2. Nomenclature of Coordination Compounds

The naming of coordination compounds follows specific rules established by the International Union of Pure and Applied Chemistry (IUPAC). The basic rules include:

Cation Naming: The cation is named first, followed by the anion. If the complex is a cation, the metal's name is used as is. If it is an anion, the metal's name is modified by adding the suffix "-ate."

Ligand Naming: Ligands are named before the metal. Anionic ligands typically end in "-o," while neutral ligands retain their names (e.g., water as "aqua," ammonia as "ammine").

Multiplicity of Ligands: The number of each type of ligand is indicated by prefixes: mono-, di-, tri-, tetra-, penta-, and hexa-.

Example: The complex [Cu(NH3)4]SO4[Cu(NH_3)_4]SO_4[Cu(NH3)4]SO4 is named tetraamminecopper(II) sulfate.

3. Geometric and Stereoisomerism in Coordination Compounds

The spatial arrangement of ligands around the central metal ion can lead to geometric and stereoisomerism in coordination compounds:

a. Geometric Isomerism

Geometric isomerism occurs in coordination complexes with different spatial arrangements of ligands. This is particularly prevalent in octahedral and square planar geometries:

Octahedral Complexes: Ligands can be arranged in different ways (facial or meridional) around the central metal. For example, [CoCl3(en)][CoCl_3(en)][CoCl3(en)] can exhibit facial (fac) and meridional (mer) isomers depending on the arrangement of chloride and ethylenediamine ligands.

Square Planar Complexes: In complexes like [NiCl2(PPh3)2][NiCl_2(PPh_3)_2][NiCl2(PPh3)2], the arrangement of ligands can create distinct isomers based on the positioning of ligands relative to each other.

b. Stereoisomerism

Stereoisomerism arises when ligands can adopt different spatial orientations without altering the connectivity of the atoms. This is prominent in complexes with bidentate ligands, leading to isomers that differ in the arrangement of the ligands.

4. Properties of Coordination Compounds

Coordination compounds exhibit unique physical and chemical properties due to their specific geometries and ligand interactions. Key properties include:

a. Color

The color of coordination complexes arises from the absorption of visible light, which excites electrons in the d-orbitals of the central metal ion. The specific colors observed depend on the nature of the metal, its oxidation state, and the type of ligands attached.

b. Magnetism

Coordination compounds can be paramagnetic or diamagnetic, depending on the presence of unpaired electrons in the d-orbitals. Transition metal complexes with unpaired electrons exhibit paramagnetism, while those with all electrons paired are diamagnetic.

c. Solubility and Stability

The solubility of coordination compounds in different solvents is influenced by the nature of the ligands and the central metal ion. Stability constants (K_stability) can be used to quantify the stability of coordination complexes, reflecting how likely they are to remain intact in solution.

5. Reactions of Coordination Compounds

Coordination compounds undergo various reactions, including:

a. Ligand Substitution Reactions

In these reactions, one ligand is replaced by another, altering the composition of the complex. For example, the substitution of water in the aqua complex [Co(H2O)6]3+[Co(H_2O)_6]^{3+}[Co(H2O)6]3+ with ammonia to form [Co(NH3)6]3+[Co(NH_3)_6]^{3+}[Co(NH3)6]3+ is a common reaction in coordination chemistry.

b. Redox Reactions

Many coordination complexes can participate in oxidation-reduction reactions, where the oxidation state of the central metal changes. This is crucial in catalytic processes and biological functions, such as electron transport in enzymes.

6. Applications of Coordination Chemistry

Coordination chemistry has numerous practical applications, including:

a. Catalysis

Coordination complexes are widely used as catalysts in various industrial processes, such as hydrogenation, polymerization, and oxidation reactions. Transition metal catalysts can enhance reaction rates and selectivity.

b. Biological Systems

Metal ions, such as iron in hemoglobin and copper in ceruloplasmin, play vital roles in biological systems. Understanding coordination chemistry is essential for elucidating metal ion functions in enzymatic processes and electron transport.

c. Material Science

Coordination compounds are utilized in the design of advanced materials, including sensors, catalysts, and drug delivery systems. Their tunable properties allow for the development of materials with specific functionalities.

d. Medical Applications

Coordination complexes are increasingly explored in medicinal chemistry for their potential use as therapeutic agents. For example, cisplatin is a platinum-based coordination complex used in cancer treatment.

Coordination chemistry and the study of coordination complexes are fundamental aspects of inorganic chemistry. The unique properties and behaviors of these compounds arise from the interactions between metal ions and ligands, leading to a wide range of applications in catalysis, biology, materials science, and medicine. Understanding the principles of coordination chemistry provides valuable insights into the nature of chemical bonding and the reactivity of inorganic compounds, making it an essential area of study for chemists and researchers across various disciplines. As new discoveries continue to emerge in this field, coordination chemistry remains a vibrant and critical area of research in the advancement of science and technology.

Transition Metals and Their Compounds

Transition metals are a group of elements found in the central block of the periodic table (groups 3 to 12). They are characterized by their ability to form variable oxidation states, a wide range of coordination complexes, and complex formation with ligands. Transition metals are integral to many chemical processes and are crucial for various industrial applications, biological functions, and materials science.

1. Definition and Characteristics of Transition Metals

Transition metals include elements such as iron (Fe), copper (Cu), nickel (Ni), and chromium (Cr). They share certain common characteristics:

Partially Filled d-Orbitals: Transition metals have partially filled d-orbitals in their atoms, which contribute to their unique chemical properties.

Variable Oxidation States: Transition metals can exhibit multiple oxidation states due to the relatively close energies of their s and d orbitals. For example, iron can exist as Fe^{2+} or Fe^{3+}.

Formation of Colored Compounds: Many transition metal compounds are colored due to d-d electron transitions, which can be observed in coordination complexes.

Catalytic Properties: Transition metals often serve as catalysts in various chemical reactions due to their ability to readily change oxidation states and form stable complexes with reactants.

2. Common Transition Metals and Their Compounds

Several transition metals are notable for their practical applications and biological significance:

a. Iron (Fe)

Oxidation States: Iron commonly exhibits oxidation states of +2 (ferrous, Fe^{2+}) and +3 (ferric, Fe^{3+}).

Compounds: Iron(III) oxide (Fe_2O_3) is a common rust component, while iron(II) sulfate ($FeSO_4$) is used in water treatment and as a dietary supplement.

Biological Importance: Iron is essential for hemoglobin in blood, facilitating oxygen transport in organisms.

b. Copper (Cu)

Oxidation States: Copper typically exists in +1 (cuprous, Cu^{+}) and +2 (cupric, Cu^{2+}) states.

Compounds: Copper(II) sulfate ($CuSO_4$) is used in agriculture and as a fungicide. Copper(I) oxide (Cu_2O) is utilized in ceramics and as a pigment.

Biological Role: Copper is a vital cofactor for several enzymes, including cytochrome c oxidase, which is involved in cellular respiration.

c. Nickel (Ni)

Oxidation States: Nickel commonly shows oxidation states of +2 (nickel(II)) and +3 (nickel(III)), although +2 is more stable.

Compounds: Nickel(II) sulfate ($NiSO_4$) is used in electroplating, while nickel(II) hydroxide ($Ni(OH)_2$) is utilized in batteries.

Catalytic Applications: Nickel is used as a catalyst in hydrogenation reactions in organic chemistry.

3. Coordination Compounds of Transition Metals

Transition metals readily form coordination compounds with ligands, leading to diverse geometries and reactivities. The nature of the ligands and the oxidation state of the metal influence the properties of these complexes.

a. Types of Ligands

Monodentate Ligands: Ligands that bind to the metal at a single site (e.g., water, ammonia).

Bidentate Ligands: Ligands that can form two bonds with the metal (e.g., ethylenediamine, oxalate).

Polydentate Ligands: Ligands that can bind at multiple sites, such as EDTA, which can form stable complexes with transition metals.

b. Geometric and Stereoisomerism

The geometry of coordination complexes often leads to isomerism:

Octahedral Complexes: Common for six-coordinate complexes, exhibiting facial and meridional isomers.

Square Planar Complexes: Found in certain d^8 metals (e.g., platinum, nickel), can exhibit cis and trans isomers.

4. Applications of Transition Metal Compounds

Transition metals and their compounds have vast applications across various fields:

a. Catalysis

Transition metals are pivotal in catalyzing chemical reactions due to their ability to change oxidation states and coordinate with reactants. They are used in:

Heterogeneous Catalysis: Transition metals like platinum and palladium are used in catalytic converters to reduce automotive emissions.

Homogeneous Catalysis: Transition metal complexes facilitate reactions in solution, such as the Haber process for ammonia synthesis using iron catalysts.

b. Biological Functions

Transition metals play crucial roles in biological systems:

Enzymes: Many enzymes require transition metal ions (e.g., iron, copper, manganese) as cofactors to facilitate biochemical reactions.

Oxygen Transport: Hemoglobin (iron-containing) and myoglobin transport oxygen in blood and muscle tissues, respectively.

c. Materials Science

Transition metals are used in the development of advanced materials, including:

Alloys: Transition metals like nickel, chromium, and cobalt are used to enhance the properties of steel and other alloys.

Magnetic Materials: Transition metal compounds are crucial in the production of magnetic materials for electronics and data storage.

5. Environmental and Industrial Impact

The extraction and processing of transition metals can significantly impact the environment. Sustainable practices are being developed to mitigate these effects:

Recycling: Transition metals are often recycled from electronic waste and industrial byproducts to reduce environmental strain.

Green Chemistry: The use of environmentally friendly methods for synthesizing and utilizing transition metal compounds is becoming a focus in chemical research.

Transition metals and their compounds are fundamental to both inorganic chemistry and numerous practical applications. Their unique properties, including variable oxidation states, the ability to form complex coordination compounds, and catalytic efficiency, make them invaluable across various industries. Understanding the behavior of transition metals not only advances scientific knowledge but also enhances our ability to address contemporary challenges in technology, health, and environmental sustainability. The ongoing study of transition metals will continue to reveal their importance and versatility, paving the way for innovative applications in the future.

Main Group Chemistry: Properties and Reactions

Main group elements are those found in Groups 1, 2, and 13 through 18 of the periodic table. These elements include metals, nonmetals, and metalloids and are characterized by their varied properties and reactivities. This chapter explores the key properties, trends, and reactions of the main group elements, emphasizing their significance in chemical processes and practical applications.

1. Overview of Main Group Elements

Main group elements can be categorized into two primary categories:

Representative Elements: These include Groups 1 (alkali metals), 2 (alkaline earth metals), and Groups 13-18 (p-block elements).

Noble Gases: Group 18 consists of the noble gases, which are characterized by their full valence electron shells and low reactivity.

2. Properties of Main Group Elements

a. Alkali Metals (Group 1)

Characteristics: Alkali metals (lithium, sodium, potassium, rubidium, cesium, and francium) are soft, highly reactive metals with one valence electron.

Physical Properties: They have low melting and boiling points compared to other metals and exhibit low densities. For instance, lithium is the lightest metal, while potassium floats on water.

Reactivity: Alkali metals react vigorously with water, producing hydroxides and hydrogen gas. For example:

$$2\mathrm{Na} + 2\mathrm{H_2O} \rightarrow 2\mathrm{NaOH} + \mathrm{H_2} \uparrow$$

Compounds: Alkali metal compounds, such as sodium chloride (table salt) and lithium carbonate, are common in nature.

b. Alkaline Earth Metals (Group 2)

Characteristics: Alkaline earth metals (beryllium, magnesium, calcium, strontium, barium, and radium) have two valence electrons and are less reactive than alkali metals.

Physical Properties: These metals are harder than alkali metals and have higher melting points and densities.

Reactivity: Alkaline earth metals react with water, but less vigorously than alkali metals. For example, magnesium reacts with water at elevated temperatures:

$$\mathrm{Mg} + 2\mathrm{H_2O} \rightarrow \mathrm{Mg(OH)_2} + \mathrm{H_2} \uparrow$$

Compounds: They form ionic compounds, such as calcium carbonate (found in limestone) and magnesium sulfate (Epsom salt).

c. P-Block Elements (Groups 13-16)

Group 13: This group includes aluminum, gallium, indium, and thallium. Aluminum is notable for its lightweight and corrosion resistance. It reacts with acids and bases, forming aluminum salts.

Group 14: Carbon, silicon, germanium, tin, and lead make up this group. Carbon, with its ability to form diverse organic compounds, is fundamental to life. Silicon is essential in technology, especially in semiconductors.

Group 15: This group consists of nitrogen, phosphorus, arsenic, antimony, and bismuth. Nitrogen is vital for biological processes, while phosphorus is a key component of DNA and ATP.

Group 16: Oxygen, sulfur, selenium, tellurium, and polonium are included in this group. Oxygen is crucial for respiration, while sulfur compounds are essential in many industrial processes.

d. Noble Gases (Group 18)

Characteristics: Noble gases (helium, neon, argon, krypton, xenon, and radon) have full valence electron shells, making them highly stable and largely inert.

Physical Properties: They are colorless, odorless, and exist as monatomic gases under standard conditions.

Applications: Noble gases are used in lighting (neon lights), welding (argon), and as inert environments for chemical reactions.

3. Periodic Trends in Main Group Elements

Several periodic trends are observed among the main group elements:

a. Atomic Size

Atomic size generally increases down a group due to the addition of electron shells. For instance, lithium is smaller than sodium.

Across a period, atomic size decreases from left to right as increased nuclear charge pulls electrons closer to the nucleus.

b. Ionization Energy

Ionization energy, the energy required to remove an electron, generally decreases down a group and increases across a period.

For example, it is easier to remove an electron from potassium than from magnesium.

c. Electronegativity

Electronegativity, the tendency of an atom to attract electrons in a bond, also varies with periodic trends. It generally increases across a period and decreases down a group.

Fluorine is the most electronegative element, while elements like cesium have low electronegativity.

4. Reactions of Main Group Elements

a. Formation of Ions

Main group metals typically lose electrons to form cations. For instance, sodium loses one electron to form Na^+.

Nonmetals tend to gain electrons, forming anions. Chlorine gains an electron to become Cl^-.

b. Acid-Base Reactions

Many main group elements participate in acid-base reactions. For example, sodium hydroxide (NaOH), a base formed from sodium, reacts with hydrochloric acid (HCl) to produce water and sodium chloride:

$$\mathrm{NaOH} + \mathrm{HCl} \rightarrow \mathrm{NaCl} + \mathrm{H_2O}$$

c. Redox Reactions

Redox reactions, involving the transfer of electrons, are common in main group chemistry. For example, the reaction of magnesium with oxygen to form magnesium oxide illustrates this process:

$$2Mg + O_2 \rightarrow 2MgO$$

d. Formation of Compounds

Main group elements can form a variety of compounds, both ionic and covalent. For example, sodium chloride (ionic) and carbon dioxide (covalent) are well-known compounds resulting from main group elements.

5. Practical Applications of Main Group Chemistry

Main group chemistry is significant in various fields:

Materials Science: Aluminum, a main group element, is widely used in construction and manufacturing due to its strength and lightweight properties.

Biological Importance: Nitrogen and phosphorus are crucial for life, forming the backbone of DNA and RNA, and are essential in fertilizers for agriculture.

Industrial Applications: Compounds like sulfuric acid (H_2SO_4), produced from sulfur, are vital in chemical manufacturing, including the production of fertilizers, detergents, and other chemicals.

Main group chemistry encompasses a wide range of elements with diverse properties and reactions. Understanding these elements and their behaviors is fundamental for fields such as chemistry, biology, materials science, and environmental science. The periodic trends observed in main group elements help predict their reactivities and interactions, making them integral to both theoretical understanding and practical applications in everyday life.

Bioinorganic Chemistry

Bioinorganic chemistry is a specialized branch of chemistry that focuses on the role of inorganic elements and compounds in biological systems. This field bridges the gap between inorganic chemistry and biochemistry, exploring how metal ions and inorganic molecules contribute to the structure and function of biological molecules. The following sections will delve into the fundamental concepts of bioinorganic chemistry, the importance of metal ions in biological processes, examples of metalloproteins and metalloenzymes, and their applications in medicine and biotechnology.

1. Overview of Bioinorganic Chemistry

Bioinorganic chemistry examines the interactions between inorganic substances and biological systems. It investigates how metal ions participate in biological functions, the nature of metalloproteins, and the mechanisms through which metals influence biological activity. Understanding bioinorganic chemistry is essential for elucidating biochemical pathways, developing drugs, and creating biomimetic materials.

2. Role of Metal Ions in Biological Systems

Metal ions play crucial roles in various biological processes. Their unique properties, such as variable oxidation states and the ability to form coordination complexes, make them indispensable in biochemistry.

a. Metals as Cofactors

Many enzymes require metal ions as cofactors for their activity. These metal ions can stabilize charged intermediates, facilitate electron transfer, or participate directly in catalytic mechanisms. Common metal cofactors include:

Iron (Fe): Essential for hemoglobin and myoglobin, where it binds oxygen. Iron is also a key component of cytochromes, involved in electron transport in cellular respiration.

Copper (Cu): Found in enzymes such as cytochrome c oxidase and superoxide dismutase, copper is vital for electron transfer and oxidative stress response.

Zinc (Zn): Functions as a structural cofactor in enzymes like carbonic anhydrase and DNA polymerases, zinc plays a role in catalysis and stabilization of protein structures.

Magnesium (Mg): Involved in many enzymatic reactions, magnesium is a cofactor for ATP-dependent processes and is crucial for DNA and RNA synthesis.

b. Metal Ions and Electron Transfer

Metal ions are critical in electron transfer reactions, particularly in redox processes. They can easily switch between different oxidation states, allowing them to accept or donate electrons during biochemical reactions. This property is essential in processes such as:

Photosynthesis: The chlorophyll molecule contains magnesium ions, which are crucial for capturing light energy and facilitating electron transfer during the light-dependent reactions of photosynthesis.

Cellular Respiration: Transition metals such as iron and copper play significant roles in the electron transport chain, where they facilitate the transfer of electrons, leading to the production of ATP.

3. Metalloproteins and Metalloenzymes

Metalloproteins are proteins that contain metal ions as part of their structure, while metalloenzymes are enzymes that contain metal ions as essential cofactors for their catalytic activity.

a. Hemoglobin and Myoglobin

Hemoglobin: A metalloprotein found in red blood cells, hemoglobin contains iron ions that bind oxygen, enabling the transport of oxygen from the lungs to tissues. The iron in hemoglobin can alternate between Fe^{2+} (deoxymyoglobin) and Fe^{3+} (metmyoglobin) states, influencing its oxygen-binding capacity.

Myoglobin: Similar to hemoglobin, myoglobin is found in muscle tissues and serves to store oxygen. Its high affinity for oxygen allows it to efficiently release oxygen during muscle contraction.

b. Cytochrome P450

Cytochrome P450 is a family of heme-containing enzymes involved in the metabolism of various substrates, including drugs and toxins. The iron in the heme group undergoes redox reactions, facilitating the oxidation of organic compounds. This enzyme is crucial for drug metabolism in the liver.

c. Carbonic Anhydrase

This enzyme, which contains a zinc ion at its active site, catalyzes the reversible reaction between carbon dioxide and water to form bicarbonate. Carbonic anhydrase plays a critical role in regulating pH and carbon dioxide levels in blood and tissues.

4. Metals in Medicine

Bioinorganic chemistry has significant implications in medicine, particularly in the development of metal-based therapeutics.

a. Chemotherapy

Cisplatin: One of the most well-known metal-based drugs, cisplatin contains platinum and is used in the treatment of various cancers. It works by forming DNA cross-links, preventing cancer cells from dividing and leading to apoptosis.

b. Diagnostic Agents

MRI Contrast Agents: Gadolinium-based compounds are commonly used as contrast agents in magnetic resonance imaging (MRI). Gadolinium's paramagnetic properties enhance image contrast, allowing for better visualization of tissues.

c. Antimicrobial Agents

Silver Compounds: Silver ions exhibit antimicrobial properties and are used in wound dressings and coatings for medical devices to prevent infections.

5. Applications in Biotechnology

Bioinorganic chemistry also finds applications in biotechnology, including the development of biosensors and biomimetic materials.

a. Biosensors

Metalloproteins and metalloenzymes can be utilized in biosensors to detect specific biological molecules. For instance, glucose oxidase, containing a flavin cofactor and metal ions, is used to measure glucose levels in diabetic patients.

b. Biomimetic Catalysts

Research in bioinorganic chemistry aims to design synthetic catalysts that mimic the function of natural enzymes. These biomimetic catalysts can be utilized in industrial processes for more sustainable and efficient chemical transformations.

6. Challenges and Future Directions

Despite the advancements in bioinorganic chemistry, several challenges remain. Understanding the precise mechanisms by which metal ions influence biological processes is crucial for developing targeted therapies and interventions. Ongoing research aims to:

Unravel the complexities of metalloproteins and metalloenzymes to enhance our understanding of their functions and interactions.

Develop new metal-based drugs with improved efficacy and reduced side effects.

Investigate the role of metal ions in diseases, such as neurodegenerative disorders linked to metal homeostasis.

Bioinorganic chemistry is a vital field that enhances our understanding of the interactions between inorganic elements and biological systems. By studying the roles of metal ions in biological processes, researchers can develop innovative applications in medicine, biotechnology, and environmental science. The integration of bioinorganic chemistry into various scientific disciplines highlights its importance in addressing contemporary challenges in health and technology. As research continues to advance, bioinorganic chemistry will play an increasingly critical role in shaping the future of science and medicine.

Chapter 12: Environmental and Industrial Chemistry

Chemistry in the Environment: Pollution and its Effects

Environmental chemistry is a sub-discipline of chemistry that focuses on the chemical processes occurring in the environment, the effects of chemical substances on ecosystems, and the impact of human activities on natural systems. In this chapter, we will discuss the various forms of pollution, their sources, and the significant effects they have on human health, ecosystems, and the environment. We will also explore the role of chemistry in mitigating pollution and promoting sustainability.

1. Understanding Pollution

Pollution is the introduction of harmful substances or contaminants into the natural environment, resulting in adverse effects on ecosystems, human health, and the overall quality of life. Pollutants can be classified into several categories based on their origin and nature:

a. Types of Pollution

Air Pollution: The presence of harmful substances in the atmosphere, including particulate matter, gases, and vapors. Common air pollutants include sulfur dioxide (SO_2), nitrogen oxides ($NO_□$),carbon monoxide (CO), volatile organic compounds (VOCs), and ozone (O_3).

Water Pollution: The contamination of water bodies such as rivers, lakes, and oceans by harmful chemicals, pathogens, and debris. Water pollutants include heavy metals (e.g., lead, mercury), nutrients (e.g., nitrates and phosphates), organic pollutants (e.g., pesticides, pharmaceuticals), and pathogens.

Soil Pollution: The degradation of soil quality due to the presence of toxic substances, often resulting from industrial activities, agricultural practices, and waste disposal. Soil pollutants can include heavy metals, pesticides, herbicides, and petroleum hydrocarbons.

Noise Pollution: Although not a chemical pollutant per se, noise pollution is considered an environmental issue caused by excessive noise from transportation, industrial activities, and urban areas that can adversely affect human health and wildlife.

Light Pollution: The excessive or misdirected artificial light in urban areas, which can disrupt ecosystems, affect human sleep patterns, and hinder astronomical observations.

2. Sources of Pollution

Pollutants can originate from both natural processes and human activities.

a. Natural Sources

Natural sources of pollution include volcanic eruptions, wildfires, and dust storms. These events release gases, particulate matter, and other substances into the environment, which can impact air quality and climate.

b. Anthropogenic Sources

Human activities are the primary contributors to environmental pollution. Key sources include:

Industrial Processes: Factories and manufacturing plants often release harmful substances into the air, water, and soil through emissions, effluents, and waste disposal. Chemical manufacturing, mining, and energy production are significant contributors.

Agricultural Practices: The use of fertilizers, pesticides, and herbicides in agriculture leads to nutrient runoff, soil degradation, and contamination of water sources. Excess nitrogen and phosphorus from fertilizers can cause algal blooms in water bodies.

Transportation: Vehicles emit various pollutants, including carbon monoxide, nitrogen oxides, particulate matter, and hydrocarbons. Transportation is a significant source of urban air pollution.

Waste Disposal: Improper waste management, including landfill overflow, incineration, and illegal dumping, releases harmful chemicals and toxins into the environment. Leachate from landfills can contaminate groundwater.

3. Effects of Pollution on Human Health

Pollution has significant adverse effects on human health, causing both acute and chronic conditions.

a. Air Pollution and Health

Exposure to air pollutants can lead to respiratory diseases (e.g., asthma, chronic obstructive pulmonary disease), cardiovascular issues, and even cancer. Vulnerable populations, such as children, the elderly, and individuals with pre-existing conditions, are at greater risk.

Particulate Matter (PM): Fine particles can penetrate deep into the lungs, leading to inflammation and exacerbating respiratory diseases.

Ground-Level Ozone: Formed by chemical reactions between volatile organic compounds and sunlight, ozone can cause respiratory problems and reduce lung function.

b. Water Pollution and Health

Contaminated water can transmit diseases and lead to various health issues. Waterborne diseases caused by pathogens (e.g., bacteria, viruses, parasites) can lead to gastrointestinal illnesses and more severe conditions.

Heavy Metals: Exposure to heavy metals like lead and mercury can cause neurological damage, developmental issues, and kidney damage.

Pharmaceuticals and Personal Care Products (PPCPs): Residues of pharmaceuticals in water bodies can disrupt endocrine systems in wildlife and humans.

c. Soil Pollution and Health

Soil contamination affects food safety and agricultural productivity. Pollutants can accumulate in crops, leading to exposure through the food chain.

Pesticides: Residues of pesticides in food can pose health risks, including neurotoxic effects and potential carcinogenicity.

4. Effects of Pollution on Ecosystems

Pollution not only impacts human health but also disrupts natural ecosystems and biodiversity.

a. Air Pollution and Ecosystems

Air pollutants can affect plant growth and biodiversity. For instance, sulfur dioxide can lead to acid rain, which harms forests, aquatic systems, and soil health.

Acid Rain: Caused by the emission of sulfur dioxide and nitrogen oxides, acid rain can lower the pH of soil and water bodies, harming aquatic life and plant species.

b. Water Pollution and Ecosystems

Nutrient pollution can lead to eutrophication, where excessive nutrients stimulate algal blooms, depleting oxygen levels and harming aquatic organisms.

Hypoxia: The depletion of oxygen in water bodies due to algal blooms can lead to fish kills and loss of aquatic biodiversity.

c. Soil Pollution and Ecosystems

Soil contaminants can adversely affect soil health, reducing its fertility and ability to support plant life. Polluted soils can lead to diminished agricultural yields and loss of habitat for various organisms.

5. Mitigating Pollution: The Role of Chemistry

Chemistry plays a crucial role in developing methods for pollution prevention, remediation, and sustainable practices.

a. Green Chemistry

Green chemistry focuses on designing chemical processes and products that minimize the use and generation of hazardous substances. Principles include:

Waste Minimization: Designing processes to reduce waste generation.

Safer Solvents: Utilizing less toxic or non-toxic solvents in chemical reactions.

Renewable Feedstocks: Using renewable resources instead of non-renewable materials.

b. Pollution Remediation

Various chemical methods are employed to remediate contaminated environments, including:

Bioremediation: Using microorganisms to degrade or remove pollutants from soil and water.

Chemical Treatments: Employing chemical agents to neutralize or extract contaminants (e.g., using activated carbon to adsorb pollutants).

c. Environmental Monitoring

Chemistry is vital in developing analytical methods for monitoring pollutants in air, water, and soil. These methods help assess the effectiveness of pollution control measures and guide regulatory decisions.

6. Legislation and Policy

Effective pollution management requires robust environmental regulations and policies. Governments and international organizations establish guidelines and standards to limit pollutant emissions and protect public health and ecosystems.

Clean Air Act: In the United States, this legislation regulates air emissions and sets standards for air quality to protect human health and the environment.

Water Quality Standards: Regulations ensure that water bodies meet specific quality criteria to protect aquatic life and human health.

7. Public Awareness and Action

Raising public awareness about pollution and its effects is essential for promoting environmental stewardship. Education can empower individuals and communities to take action toward reducing their environmental footprint.

a. Community Initiatives

Community-led initiatives can focus on reducing waste, promoting recycling, and advocating for cleaner technologies. Such grassroots efforts can have a significant impact on local pollution levels.

b. Sustainable Practices

Individuals can adopt sustainable practices, such as using public transportation, reducing energy consumption, and supporting local and organic products, to minimize their environmental impact.

Pollution poses significant challenges to human health, ecosystems, and the environment. Understanding the sources and effects of pollution is crucial for developing effective strategies to mitigate its impact. Chemistry plays a vital role in this endeavor, offering solutions for pollution prevention, remediation, and sustainable practices. By integrating scientific knowledge with community action and policy measures, we can work toward a cleaner, healthier environment for future generations. As we face the growing challenges of pollution, collaboration among scientists, policymakers, and the public is essential to achieve sustainable solutions and protect our planet.

Green Chemistry and Sustainable Practices

Green chemistry is a revolutionary approach to chemical research and industrial processes aimed at minimizing the environmental impact of chemical production and use. It seeks to design chemical products and processes that reduce or eliminate the generation of hazardous substances, promoting sustainability in the field of chemistry. This chapter explores the principles of green chemistry, its applications in various industries, and the importance of sustainable practices in addressing environmental challenges.

1. Understanding Green Chemistry

Green chemistry, often referred to as sustainable chemistry, is based on the concept of reducing the ecological footprint of chemical practices. It emphasizes designing safer chemicals and processes, which not only benefit human health but also protect the environment.

a. Core Principles of Green Chemistry

Green chemistry is guided by twelve principles established by Paul Anastas and John Warner. These principles serve as a framework for developing environmentally friendly chemical processes:

Prevention: Minimizing waste generation is preferable to treating or cleaning up waste after it has been created.

Atom Economy: Synthetic methods should be designed to maximize the incorporation of all materials used in the process into the final product, reducing waste.

Less Hazardous Chemical Syntheses: Synthetic methods should be designed to use and generate substances that possess little or no toxicity to human health and the environment.

Designing Safer Chemicals: Chemical products should be designed to preserve efficacy while minimizing toxicity.

Solvent-Free Processes: If possible, solvents should be avoided; if their use is necessary, they should be innocuous and environmentally benign.

Energy Efficiency: Energy requirements should be minimized, and processes should be conducted at ambient temperature and pressure whenever possible.

Renewable Feedstocks: Raw materials should be renewable rather than depleting whenever technically and economically practicable.

Reduce Derivatives: Unnecessary derivatization (protection, blocking, or modification of functional groups) should be minimized or avoided if possible.

Catalysis: Catalytic reagents (as selective as possible) are superior to stoichiometric reagents.

Design for Degradation: Chemical products should be designed so that at the end of their function, they break down into innocuous degradation products that do not persist in the environment.

Real-Time Analysis for Pollution Prevention: Analytical methodologies need to be further developed to allow for real-time monitoring and control during syntheses.

Inherently Safer Chemistry for Accident Prevention: Substances and the form of substances used in a chemical process should be chosen to minimize the potential for chemical accidents, including releases, explosions, and fires.

2. Applications of Green Chemistry

The principles of green chemistry can be applied across various sectors, leading to sustainable practices that address environmental challenges. Here are some notable applications:

a. Industrial Processes

Green Synthesis: Industries are increasingly adopting green synthesis techniques to produce chemicals. For example, the pharmaceutical industry is moving toward using more sustainable methods for drug synthesis, such as using biocatalysts to reduce the environmental impact of chemical reactions.

Biotechnology: The use of microbial fermentation processes in the production of biofuels, enzymes, and bioplastics exemplifies green chemistry applications. These processes can replace traditional petrochemical processes, reducing reliance on fossil fuels.

Material Science: Development of biodegradable materials, such as polylactic acid (PLA), from renewable resources exemplifies the integration of green chemistry principles in material design.

b. Agriculture

Biopesticides and Biofertilizers: The development of biopesticides and biofertilizers as alternatives to chemical pesticides and fertilizers supports sustainable agriculture by reducing chemical runoff and promoting soil health.

Precision Agriculture: Utilizing data-driven technologies and chemical formulations in precision agriculture can minimize chemical inputs while maximizing crop yields, resulting in less environmental impact.

3. The Role of Renewable Energy

Transitioning from fossil fuels to renewable energy sources is critical for implementing sustainable practices in green chemistry. Renewable energy technologies can power chemical processes with lower emissions.

Solar Energy: Solar-powered chemical processes are being developed, utilizing solar energy for reactions that traditionally require high temperatures and pressures.

Biofuels: The production of biofuels from biomass and waste materials offers a sustainable alternative to traditional fossil fuels, reducing greenhouse gas emissions.

4. Waste Minimization and Management

Green chemistry emphasizes the importance of waste reduction and efficient resource management. Techniques for waste minimization include:

Waste Valorization: Converting waste materials into valuable products. For example, using agricultural waste to produce bioplastics or energy can transform waste into a resource.

Recycling and Reuse: Promoting recycling of chemical materials and encouraging the design of products that are easier to recycle at the end of their life cycle.

Life Cycle Assessment: Evaluating the environmental impact of a product throughout its entire life cycle, from raw material extraction to disposal, can help identify areas for improvement.

5. Education and Awareness

Educating the next generation of chemists about the principles of green chemistry is essential for fostering a culture of sustainability within the scientific community. Universities and research institutions are increasingly incorporating green chemistry into their curricula, preparing students to address environmental challenges effectively.

Training Programs: Workshops and training programs focused on green chemistry practices can empower professionals in various industries to implement sustainable methods in their work.

Public Awareness Campaigns: Promoting awareness of green chemistry principles and their benefits to the public can foster a more environmentally conscious society and encourage consumer demand for sustainable products.

6. Challenges and Barriers to Implementation

While the principles of green chemistry offer substantial benefits, several challenges hinder their widespread adoption:

Economic Factors: The initial costs associated with implementing green chemistry practices and technologies can be high, deterring companies from making the switch.

Regulatory Barriers: Existing regulations may not always support the adoption of green chemistry practices, and navigating the regulatory landscape can be complex.

Research and Development: The need for continued research and development to innovate and validate new green chemistry processes remains a priority.

Industry Resistance: Some industries may be resistant to change, especially if traditional methods are perceived as more convenient or cost-effective.

Green chemistry represents a vital approach to addressing environmental challenges by promoting sustainability in chemical processes. By adopting the principles of green chemistry, industries can minimize waste, reduce toxicity, and harness renewable resources, leading to a healthier planet. The shift toward sustainable practices requires collaboration among scientists, industries, policymakers, and the public to create a sustainable future. Continued research, education, and innovation in green chemistry will play a critical role in achieving a balance between human needs and environmental protection, paving the way for a more sustainable world.

Industrial Chemistry: Processes and Applications

Industrial chemistry is a branch of chemistry that deals with the production and use of chemical products in various industries. This field is essential for the development of materials, fuels, and chemicals that drive economic growth and improve quality of life. Understanding the processes and applications of industrial chemistry is crucial for developing sustainable practices that minimize environmental impact. This chapter delves into key processes in industrial chemistry, their applications across various sectors, and the challenges and innovations shaping the future of this field.

1. Overview of Industrial Chemistry

Industrial chemistry encompasses the large-scale production of chemicals, materials, and energy. It integrates principles of chemistry, engineering, and economics to design and optimize processes for the synthesis of chemical products. The goal is to maximize efficiency, reduce costs, and minimize environmental impact while meeting societal needs.

a. Importance of Industrial Chemistry

Economic Impact: Industrial chemistry contributes significantly to the global economy by providing essential materials for various sectors, including pharmaceuticals, agriculture, energy, and manufacturing.

Technological Advancement: Innovations in industrial chemistry lead to the development of new materials and processes that enhance productivity and sustainability.

Environmental Stewardship: Industrial chemists are increasingly focused on reducing waste and emissions, developing cleaner production methods, and creating sustainable products.

2. Key Processes in Industrial Chemistry

Industrial chemistry employs a variety of processes for the production of chemicals and materials. Some of the key processes include:

a. Synthesis

Chemical Synthesis: This involves the combination of raw materials through chemical reactions to produce desired products. It can be categorized into:

Organic Synthesis: Producing organic compounds, such as pharmaceuticals and polymers, through various reaction mechanisms.

Inorganic Synthesis: Involves the preparation of inorganic compounds, including metals, salts, and coordination complexes.

Catalysis: Catalysts are substances that accelerate chemical reactions without being consumed. Industrial processes often utilize catalysts to increase reaction rates, reduce energy requirements, and improve selectivity. Examples include:

Haber-Bosch Process: This process synthesizes ammonia from nitrogen and hydrogen, using iron catalysts under high pressure and temperature, vital for fertilizer production.

Catalytic Cracking: Used in petroleum refining to break down large hydrocarbons into gasoline and other fuels.

b. Separation Processes

Separation processes are crucial in industrial chemistry for purifying products and removing impurities. Common techniques include:

Distillation: A widely used method for separating components based on differences in boiling points. It is commonly used in the petrochemical industry to refine crude oil into fuels and chemicals.

Filtration: This physical process separates solids from liquids or gases using porous materials. It is essential in various industries, including pharmaceuticals and water treatment.

Extraction: This process involves separating a substance from a mixture using a solvent. Liquid-liquid extraction and solid-liquid extraction are common techniques used in the chemical industry to purify compounds.

Membrane Technologies: These processes utilize semi-permeable membranes to separate substances based on size, charge, or other properties. Applications include water purification and gas separation.

c. Chemical Reaction Engineering

The design and optimization of chemical reactors are central to industrial chemistry. Key considerations include:

Reactor Design: Various types of reactors are employed, such as batch, continuous, and plug flow reactors, each suited to different reaction conditions and scales.

Process Control: Monitoring and controlling reaction conditions, such as temperature, pressure, and concentration, are crucial for maximizing yield and ensuring product quality.

Scale-Up: Transitioning from laboratory-scale experiments to full-scale production involves careful consideration of kinetics, thermodynamics, and material handling.

3. Applications of Industrial Chemistry

Industrial chemistry plays a vital role in numerous sectors, providing essential materials and products. Key applications include:

a. Pharmaceuticals

Drug Development: Industrial chemistry is crucial in the synthesis and production of pharmaceuticals. Techniques such as combinatorial chemistry and high-throughput screening are used to discover and develop new drugs.

Formulation Science: Developing stable and effective formulations, including tablets, injections, and creams, requires a deep understanding of chemistry.

b. Agriculture

Fertilizers and Pesticides: The production of nitrogen-based fertilizers (e.g., urea) and biopesticides involves various chemical processes that enhance agricultural productivity and protect crops from pests.

Herbicides: The synthesis of herbicides helps in controlling weeds, improving crop yields, and ensuring food security.

c. Energy Production

Fuels: Industrial chemistry is essential for refining crude oil into gasoline, diesel, and other fuels. The production of biofuels from biomass and waste materials is also gaining importance in reducing reliance on fossil fuels.

Battery Technologies: The development of materials for batteries, including lithium-ion and solid-state batteries, is crucial for advancing energy storage technologies.

d. Materials Science

Polymers: Industrial chemistry is fundamental in the synthesis of polymers used in packaging, textiles, and construction. Innovations in polymer chemistry have led to the development of biodegradable and recyclable materials.

Composites: The production of composite materials that combine different properties for enhanced performance is a key area of industrial chemistry.

e. Environmental Applications

Water Treatment: Industrial chemistry techniques are employed in the treatment and purification of water, ensuring access to safe drinking water and minimizing environmental pollution.

Waste Management: Chemical processes are used in recycling and waste treatment to minimize landfill waste and recover valuable resources.

4. Challenges in Industrial Chemistry

While industrial chemistry offers numerous benefits, several challenges must be addressed:

a. Environmental Concerns

Pollution: Industrial processes can generate hazardous waste and emissions that harm the environment and human health. Adopting cleaner technologies and waste management practices is essential.

Resource Depletion: The reliance on finite resources, such as fossil fuels and metals, poses challenges for sustainability. Transitioning to renewable resources is crucial.

b. Economic Factors

Cost of Implementation: The initial costs of adopting new technologies and processes can be prohibitive for some industries. Finding economically viable solutions is necessary for widespread adoption.

Market Dynamics: Fluctuations in market demand and raw material prices can impact the viability of industrial chemistry processes.

5. Innovations and Future Directions

The field of industrial chemistry is constantly evolving, driven by technological advancements and societal needs. Key trends shaping the future include:

a. Sustainable Practices

Green Chemistry: The integration of green chemistry principles in industrial processes aims to minimize waste and reduce environmental impact. Innovations in sustainable synthesis and biocatalysis are on the rise.

Circular Economy: Emphasizing recycling and reuse of materials promotes sustainability and reduces resource depletion.

b. Digital Transformation

Data Analytics: The use of data analytics and artificial intelligence in optimizing processes, improving efficiency, and predicting market trends is becoming increasingly important in industrial chemistry.

Automation and Robotics: Automation in manufacturing processes enhances precision, reduces labor costs, and improves safety.

c. Collaboration and Research

Interdisciplinary Research: Collaborations between chemists, engineers, and environmental scientists are crucial for developing innovative solutions to complex challenges in industrial chemistry.

Industry Partnerships: Partnerships between academia and industry can foster innovation and lead to the commercialization of new technologies.

Industrial chemistry is a dynamic field that plays a crucial role in the production of materials and chemicals essential for modern society. By integrating sustainable practices, leveraging technological advancements, and addressing environmental concerns, industrial chemistry can contribute to a more sustainable future. As the industry continues to evolve, the commitment to innovation and collaboration will be vital in overcoming challenges and maximizing the benefits of industrial chemistry for society and the environment.

Chemistry in Medicine and Pharmaceuticals

Chemistry plays a pivotal role in the field of medicine and pharmaceuticals, influencing everything from drug discovery to formulation, production, and delivery. This chapter explores the fundamental contributions of chemistry to healthcare, focusing on how chemical principles and techniques are employed to develop effective treatments and improve patient outcomes.

1. Overview of Medicinal Chemistry

Medicinal chemistry is the discipline at the intersection of chemistry and pharmacology, dedicated to the design, development, and optimization of pharmaceutical compounds. It combines knowledge from various fields, including organic chemistry, biochemistry, and pharmacology, to create safe and effective medications.

a. Importance of Medicinal Chemistry

Drug Development: Medicinal chemistry is critical in the drug discovery process, which includes target identification, lead compound discovery, optimization, and preclinical testing.

Personalized Medicine: Advances in chemistry have led to the development of targeted therapies that tailor treatments based on individual patient profiles, improving efficacy and reducing side effects.

2. Drug Discovery Process

The journey of a new drug from conception to market involves several key stages:

a. Target Identification and Validation

Understanding Diseases: Medicinal chemists begin by understanding the biological mechanisms of diseases, identifying molecular targets (e.g., proteins, enzymes, or receptors) that play critical roles in disease progression.

Validation: Once a target is identified, its role in the disease must be validated through biological assays, ensuring that modulating this target will have a therapeutic effect.

b. Lead Compound Discovery

High-Throughput Screening (HTS): Libraries of chemical compounds are screened to identify potential lead compounds that exhibit activity against the target. HTS utilizes automated techniques to test thousands of compounds quickly.

Structure-Activity Relationship (SAR): Once potential leads are identified, medicinal chemists analyze the relationship between the chemical structure of compounds and their biological activity. This analysis guides further modifications to improve efficacy and reduce toxicity.

c. Optimization of Lead Compounds

Chemical Modification: Lead compounds undergo structural modifications to enhance their pharmacological properties, including potency, selectivity, and solubility.

Pharmacokinetics and Pharmacodynamics: Medicinal chemists assess the pharmacokinetic (how the body absorbs, distributes, metabolizes, and excretes a drug) and pharmacodynamic (the effects of the drug on the body) properties of compounds to ensure safety and effectiveness.

3. Formulation and Delivery

Once a drug has been developed, it must be formulated into a suitable dosage form for administration. This involves the combination of the active pharmaceutical ingredient (API) with excipients—inactive substances that aid in the delivery of the drug.

a. Dosage Forms

Solid Forms: Tablets and capsules are common solid dosage forms. The formulation process includes ensuring stability, bioavailability (the degree and rate at which a drug is absorbed), and patient compliance.

Liquid Forms: Solutions, syrups, and suspensions are liquid formulations designed for easier administration, especially in pediatric or geriatric patients.

Parenteral Products: Injectable formulations must be sterile and stable, requiring rigorous testing to ensure safety and efficacy.

Topical Applications: Creams, ointments, and gels are formulated for local action, targeting specific areas of the body.

b. Drug Delivery Systems

Innovations in drug delivery systems aim to enhance the bioavailability and therapeutic efficacy of drugs while minimizing side effects:

Controlled Release: Techniques that allow drugs to be released over a specified period enhance patient compliance and reduce the frequency of dosing.

Targeted Delivery: Nanotechnology and conjugated drugs can deliver medications directly to the site of action, minimizing systemic exposure and side effects.

Transdermal Patches: These provide a non-invasive method of drug delivery through the skin, offering controlled release and improved patient adherence.

4. Regulatory Considerations

The pharmaceutical industry is highly regulated to ensure the safety and efficacy of drugs before they reach the market. Key regulatory bodies include:

a. Food and Drug Administration (FDA)

In the United States, the FDA oversees the drug approval process, requiring extensive preclinical and clinical trials to evaluate a drug's safety, efficacy, and quality.

b. International Regulations

Other countries have their regulatory bodies, such as the European Medicines Agency (EMA) in Europe, which also sets stringent guidelines for drug development and approval.

5. Pharmaceutical Manufacturing

Once a drug is approved, it enters the manufacturing phase, where large-scale production is essential.

a. Synthesis and Scale-Up

The chemical synthesis of pharmaceuticals must be optimized for large-scale production, balancing cost, efficiency, and environmental impact.

b. Quality Control and Assurance

Ensuring the quality of pharmaceutical products involves rigorous testing for purity, potency, and stability throughout the manufacturing process. Good Manufacturing Practices (GMP) guidelines are followed to maintain high-quality standards.

6. Challenges in Pharmaceutical Chemistry

Despite the advancements in pharmaceutical chemistry, several challenges remain:

a. Drug Resistance

Antibiotic Resistance: The emergence of drug-resistant pathogens poses a significant challenge to public health, necessitating the discovery of new antibiotics and treatment strategies.

b. Cost of Development

High R&D Costs: The drug development process is costly and time-consuming, with many compounds failing to reach the market. Balancing innovation with affordability is crucial.

c. Safety and Side Effects

Adverse Reactions: Despite rigorous testing, some drugs can cause adverse reactions in patients, necessitating ongoing monitoring and post-marketing surveillance.

7. Future Directions in Medicinal Chemistry

The field of medicinal chemistry is evolving rapidly, driven by advancements in technology and scientific knowledge. Key trends include:

a. Biologics and Biopharmaceuticals

Monoclonal Antibodies: These targeted therapies have revolutionized the treatment of various diseases, including cancer and autoimmune disorders, by specifically targeting disease-causing molecules.

b. Gene Therapy and RNA-Based Therapeutics

CRISPR Technology: Advances in gene editing offer promising avenues for treating genetic disorders by directly modifying the underlying genetic material.

mRNA Vaccines: The rapid development of mRNA vaccines for diseases like COVID-19 highlights the potential of RNA-based therapeutics in modern medicine.

c. Sustainability in Drug Development

Green Chemistry: The integration of sustainable practices in pharmaceutical development aims to reduce waste and minimize the environmental impact of drug manufacturing.

Chemistry is fundamental to the advancement of medicine and pharmaceuticals, playing a critical role in drug discovery, development, formulation, and delivery. Through the integration of chemical principles with innovative technologies, the pharmaceutical industry continues to evolve, providing new and effective treatments for various health conditions. Addressing the challenges of drug resistance, high development costs, and safety will require ongoing collaboration between chemists, pharmacologists, regulatory bodies, and healthcare professionals. The future of medicine will increasingly rely on the insights and innovations derived from chemistry, emphasizing the importance of this field in enhancing global health outcomes.

Chapter 13: Laboratory Techniques and Safety

Basic Laboratory Equipment and Their Uses

In any chemistry laboratory, having the right equipment is essential for conducting experiments safely and effectively. This chapter focuses on the basic laboratory equipment commonly used in chemistry labs, their specific functions, and best practices for using them to ensure accurate results and safety.

1. Types of Basic Laboratory Equipment

Laboratory equipment can be broadly categorized into several groups based on their functions. Below is an overview of the most common types of laboratory equipment and their uses:

a. Measuring Equipment

Graduated Cylinder:

Use: This cylindrical container is used for measuring the volume of liquids. The markings on the side allow for precise readings.

Best Practices: Always read the measurement at eye level to avoid parallax errors, and ensure the cylinder is on a flat surface.

Pipette:

Use: A pipette is used to transfer small volumes of liquid accurately. There are different types, including graduated pipettes and volumetric pipettes.

Best Practices: Use a pipette bulb or pipette filler to avoid mouth suction, and ensure the pipette is rinsed with the solution to be measured before use.

Burette:

Use: Commonly used in titration experiments, a burette allows for the precise dispensing of variable volumes of liquid.

Best Practices: Always ensure the tap is closed before filling, and check for air bubbles in the nozzle before starting the titration.

b. Heating Equipment

Bunsen Burner:

Use: A Bunsen burner provides a controlled flame for heating substances in the laboratory.

Best Practices: Always use a heat-resistant mat under the burner, and keep flammable materials away from the flame. Adjust the air intake for a clean flame.

Hot Plate:

Use: Used for heating liquids or solids, a hot plate provides a stable surface and can maintain consistent temperatures.

Best Practices: Avoid using glassware that is not rated for high temperatures, and never leave the hot plate unattended while in use.

Oven:

Use: Laboratory ovens are used for drying or sterilizing equipment and samples.

Best Practices: Ensure the oven is preheated to the desired temperature and monitor the samples to prevent overheating.

c. Mixing and Stirring Equipment

Magnetic Stirrer:

Use: A magnetic stirrer uses a rotating magnetic field to stir solutions, often used in conjunction with a stir bar.

Best Practices: Ensure the stir bar is the correct size for the container and avoid overfilling to prevent spillage.

Vortex Mixer:

Use: This device rapidly mixes small volumes of liquids, creating a vortex that helps to combine samples.

Best Practices: Use appropriate containers to avoid spillage, and ensure the lid is securely fastened if applicable.

d. Glassware

Beaker:

Use: Beakers are used for holding, mixing, and heating liquids. They come in various sizes and have graduated markings for approximate measurements.

Best Practices: Use a beaker for rough measurements, and avoid using it for precise measurements.

Erlenmeyer Flask:

Use: This flask is ideal for mixing and heating solutions, with a wide base and narrow neck to reduce evaporation.

Best Practices: Use when swirling liquids or during titration to minimize spills.

Volumetric Flask:

Use: Designed for precise dilutions and preparation of standard solutions, volumetric flasks have a single calibrated line for accuracy.

Best Practices: Fill to the calibration line using a pipette or dropper for accuracy.

Test Tubes:

Use: Test tubes are used for holding small amounts of substances, typically during qualitative experiments.

Best Practices: Avoid using them for prolonged heating, and use a test tube holder when handling hot test tubes.

e. Separation Equipment

Separatory Funnel:

Use: Used for liquid-liquid extractions, a separatory funnel allows the separation of immiscible liquids based on their densities.

Best Practices: Always ensure that the stopcock is closed before adding liquids and vent the funnel periodically to release pressure.

Filtration Setup:

Use: Used for separating solids from liquids, this setup typically includes filter paper and a funnel.

Best Practices: Wet the filter paper before use to ensure it adheres to the funnel and prevents leaks.

f. Safety Equipment

Safety Goggles:

Use: Essential personal protective equipment (PPE) for eye protection against chemical splashes and debris.

Best Practices: Always wear safety goggles when in the lab, regardless of the experiment being conducted.

Lab Coat:

Use: Protects skin and clothing from spills and splashes.

Best Practices: Ensure that the lab coat is long-sleeved and made of a material that is resistant to chemicals.

Fume Hood:

Use: Provides ventilation and protects the user from inhaling hazardous fumes and vapors during experiments.

Best Practices: Ensure the sash is at the appropriate height and keep materials at least six inches inside the hood to maximize airflow.

2. Best Practices for Using Laboratory Equipment

Understanding the proper use of laboratory equipment is essential for achieving accurate results and maintaining safety in the lab. Here are several best practices:

a. Read Instructions Carefully

Always read the operating instructions for any equipment before use. Familiarize yourself with its specific functions and safety precautions.

b. Calibrate Equipment Regularly

Regular calibration of measuring devices (such as balances and pipettes) is essential to ensure accuracy. Follow laboratory protocols for calibration checks.

c. Maintain Cleanliness

Keep the laboratory workspace organized and clean. Regularly clean equipment after use, and store it in designated places to prevent contamination and accidents.

d. Practice Safety Protocols

Follow all safety guidelines, including wearing appropriate PPE, understanding emergency procedures, and knowing the location of safety equipment such as eyewash stations and fire extinguishers.

e. Seek Assistance When Needed

If uncertain about using any equipment, seek assistance from a knowledgeable lab technician or instructor. Proper training can prevent accidents and equipment damage.

Basic laboratory equipment is the backbone of any chemistry lab, facilitating accurate measurements, safe handling, and effective experimentation. Understanding the specific uses and best practices for each type of equipment is essential for both novice and experienced chemists. By emphasizing safety and proper technique, laboratory personnel can ensure a productive and secure working environment, ultimately leading to successful experimental outcomes. Proper training and adherence to safety protocols not only protect individuals but also contribute to the integrity and reliability of scientific research.

Techniques: Measuring, Mixing, and Heating

In the field of chemistry, mastering laboratory techniques is crucial for obtaining reliable data and ensuring safety. This chapter will discuss essential techniques for measuring, mixing, and heating substances in a laboratory setting. Each technique is fundamental to performing experiments effectively and safely, highlighting best practices and equipment involved.

1. Measuring Techniques

Measuring is a vital technique in chemistry that ensures the accuracy of experiments. Accurate measurements of mass, volume, and temperature are essential for achieving reliable results.

a. Mass Measurement

Analytical Balance:

Use: Analytical balances are used to measure mass with high precision, typically to the nearest 0.0001 grams. They are equipped with draft shields to prevent air currents from affecting measurements.

Best Practices:

Tare the balance before weighing the sample to zero out the container's weight.

Use gloves or tweezers to handle samples to avoid contamination.

Ensure the balance is on a stable, level surface away from vibrations.

Top-Loading Balance:

Use: For less precise measurements, a top-loading balance can be used, measuring to the nearest 0.01 grams.

Best Practices:

Similar to the analytical balance, tare the balance before use.

Avoid weighing extremely hot or cold samples, as this can affect accuracy.

b. Volume Measurement

Graduated Cylinder:

Use: A graduated cylinder is used for measuring the volume of liquids. It is more accurate than beakers but less precise than volumetric flasks.

Best Practices:

Place the graduated cylinder on a level surface and read the meniscus (the curved surface of the liquid) at eye level for accuracy.

Use a glass rod to help pour liquids without spills.

Pipettes:

Use: Pipettes are used for transferring small, precise volumes of liquids.

Best Practices:

Use a pipette bulb to fill the pipette, avoiding mouth suction.

Always rinse the pipette with the solution to be measured before the actual measurement.

Volumetric Flask:

Use: For preparing solutions of precise concentrations, volumetric flasks are calibrated for a specific volume.

Best Practices:

Fill the flask to the calibration line, using a dropper or pipette for final adjustments.

Stopper the flask and invert it several times to ensure thorough mixing.

c. Temperature Measurement

Thermometers:

Use: Thermometers are used to measure temperature, with digital thermometers providing quick and accurate readings.

Best Practices:

Ensure the thermometer is calibrated before use.

Insert the thermometer into the substance without touching the sides of the container.

2. Mixing Techniques

Mixing is another essential laboratory technique, often used to ensure homogeneity in solutions or to initiate reactions between substances.

a. Stirring

Magnetic Stirrer:

Use: A magnetic stirrer uses a rotating magnetic field to stir solutions without manual intervention.

Best Practices:

Ensure the stir bar is appropriately sized for the container.

Avoid overfilling the container to prevent spills during stirring.

Glass Rod:

Use: A glass rod can be used for manual stirring, especially in small containers.

Best Practices:

Use a gentle motion to avoid splashing.

Rinse the rod between uses to prevent cross-contamination.

b. Vortex Mixing

Vortex Mixer:

Use: This device is used to rapidly mix small volumes of liquids, creating a vortex in the liquid.

Best Practices:

Ensure the container is securely held in place to avoid spills.

Monitor the mixing time to prevent overheating of samples.

c. Emulsification

Homogenizer:

Use: A homogenizer is used for creating emulsions or dispersions of liquids that do not normally mix.

Best Practices:

Follow manufacturer guidelines for operating the homogenizer.

Use appropriate containers that can withstand the forces applied during homogenization.

3. Heating Techniques

Heating substances is a fundamental technique in many chemistry experiments. Understanding the various heating methods and safety protocols is crucial.

a. Bunsen Burner

Use: A Bunsen burner provides a controlled flame for heating liquids and solids in the laboratory.

Best Practices:

Adjust the air supply to achieve a clean, blue flame.

Keep flammable materials away from the burner.

Always use heat-resistant mats to protect surfaces.

b. Hot Plate

Use: A hot plate allows for the even heating of liquids and solids without an open flame.

Best Practices:

Always check the temperature settings before placing glassware on the hot plate.

Avoid using glassware that is not designed for high temperatures.

c. Oven

Use: Laboratory ovens are used for drying samples or sterilizing glassware.

Best Practices:

Preheat the oven to the desired temperature before placing items inside.

Use heat-resistant gloves when handling hot items to prevent burns.

4. Safety Considerations

Ensuring safety in the laboratory during measuring, mixing, and heating is paramount. Here are key safety practices to follow:

a. Personal Protective Equipment (PPE)

Always wear appropriate PPE, including safety goggles, gloves, and lab coats, to protect against chemical exposure.

Ensure that long hair is tied back and loose clothing is secured to avoid entanglement in equipment.

b. Proper Ventilation

Conduct experiments involving volatile substances or strong fumes in a fume hood to prevent inhalation of hazardous vapors.

c. Emergency Preparedness

Familiarize yourself with the location of safety equipment, such as eyewash stations, safety showers, and fire extinguishers.

Have a clear understanding of the emergency procedures in case of chemical spills, fires, or accidents.

Mastering the techniques of measuring, mixing, and heating is essential for conducting successful chemistry experiments. By utilizing the appropriate equipment and following best practices, laboratory personnel can ensure accuracy, efficiency, and safety in their work. Understanding these foundational techniques not only enhances experimental outcomes but also contributes to a safer laboratory environment. Continuous training and adherence to safety protocols are crucial for fostering a culture of safety in the laboratory.

Safety Protocols and Proper Lab Practices

Safety in the chemistry laboratory is of utmost importance. A well-structured approach to safety protocols and proper laboratory practices not only protects the individual conducting the experiments but also safeguards the integrity of the research and the environment. This chapter outlines essential safety measures and proper lab practices that every student and professional should adhere to while working in a chemistry lab.

1. Understanding the Importance of Safety Protocols

Safety protocols in a chemistry laboratory are designed to prevent accidents, injuries, and exposure to hazardous materials. Understanding and adhering to these protocols is crucial for several reasons:

Protection from Hazards: Chemistry laboratories often involve the use of chemicals that can be toxic, flammable, or reactive. Adhering to safety protocols minimizes the risk of exposure to these hazards.

Preventing Environmental Damage: Proper disposal and handling of chemicals prevent environmental contamination and promote sustainability.

Legal Compliance: Many countries have regulations governing laboratory safety. Following these protocols ensures compliance with legal requirements.

2. Personal Protective Equipment (PPE)

Personal Protective Equipment is essential in protecting individuals from exposure to hazardous materials. The appropriate use of PPE includes:

Safety Goggles: Always wear safety goggles to protect your eyes from chemical splashes or flying debris. Goggles should fit snugly and be resistant to the chemicals being handled.

Lab Coats: Lab coats made from flame-resistant materials should be worn at all times. They protect skin and clothing from spills and splashes.

Gloves: Use appropriate gloves made of materials resistant to the chemicals being handled. Latex gloves may not be suitable for all chemicals; therefore, select gloves based on the chemical's properties.

Face Shields: In situations involving high-risk procedures or chemicals, face shields provide additional protection for the face.

Footwear: Closed-toe shoes made of durable material should be worn to protect feet from spills, broken glass, or heavy objects.

3. Proper Laboratory Practices

In addition to wearing PPE, implementing proper laboratory practices is crucial for maintaining a safe working environment. Key practices include:

a. Familiarization with the Laboratory

Know the Layout: Familiarize yourself with the laboratory layout, including exits, eyewash stations, safety showers, and fire extinguishers.

Read Safety Data Sheets (SDS): Before using any chemical, review the Safety Data Sheets to understand its hazards, safe handling practices, and emergency measures.

b. Handling Chemicals Safely

Labeling: Ensure all chemicals are clearly labeled with their names, concentrations, and hazard symbols. Never use unlabeled chemicals.

Avoiding Direct Contact: Always handle chemicals with care. Use tools like tongs or spatulas to avoid direct contact.

Working with Volatile Chemicals: Conduct experiments involving volatile substances in a fume hood to avoid inhalation of fumes.

c. Conducting Experiments Responsibly

Follow Protocols: Adhere to written procedures and protocols when conducting experiments. Deviating from established protocols can lead to accidents.

Work in Pairs: Whenever possible, work with a partner in the lab. This practice ensures that help is available in case of an emergency.

Stay Focused: Avoid distractions such as eating, drinking, or using personal electronic devices in the laboratory.

4. Emergency Preparedness

Despite the best safety practices, emergencies can occur. Preparedness is key to minimizing the impact of such situations. Key aspects of emergency preparedness include:

a. Emergency Equipment

Eyewash Stations: Know the location of eyewash stations and how to use them. They should be accessible within 10 seconds from any point in the lab.

Safety Showers: Safety showers are essential for decontaminating the body in case of chemical exposure. Ensure you know how to operate them.

Fire Extinguishers: Familiarize yourself with the types of fire extinguishers available in the lab (e.g., foam, dry powder, CO2) and their appropriate uses.

b. Incident Reporting

Report Incidents: All accidents, spills, or near-misses should be reported immediately to the laboratory supervisor. This practice ensures proper documentation and follow-up.

Emergency Procedures: Be aware of the specific emergency procedures for various scenarios (chemical spills, fires, medical emergencies) and ensure everyone in the lab is trained in these procedures.

5. Proper Disposal of Chemicals

Disposing of chemicals safely is critical for maintaining laboratory safety and environmental integrity. Key disposal practices include:

Chemical Waste Segregation: Separate chemical waste based on compatibility. For example, acids should not be mixed with bases, and organic solvents should be kept apart from aqueous waste.

Use of Designated Containers: Always dispose of waste in designated containers that are clearly labeled. Do not use regular trash bins for chemical waste.

Follow Institutional Guidelines: Each institution has specific protocols for chemical waste disposal. Familiarize yourself with these guidelines and follow them diligently.

6. Regular Training and Updates

Safety training should be an ongoing process in any laboratory setting. Regular training sessions on safety protocols and best practices help reinforce the importance of safety and keep everyone updated on new regulations and procedures.

Safety protocols and proper laboratory practices are fundamental to a successful and secure chemistry laboratory experience. By using appropriate personal protective equipment, following established procedures, preparing for emergencies, and ensuring proper chemical disposal, individuals can create a safe and efficient laboratory environment. Continuous education and adherence to safety standards are essential for fostering a culture of safety in the lab. Ultimately, a proactive approach to safety will minimize risks and enhance the overall quality of laboratory work.

Data Analysis and Interpretation

In any scientific endeavor, the collection, analysis, and interpretation of data are fundamental processes that determine the validity and reliability of experimental results. This chapter delves into the essential aspects of data analysis and interpretation in the context of a chemistry laboratory, emphasizing methods, tools, and best practices to ensure accurate conclusions can be drawn from experimental findings.

1. The Importance of Data Analysis

Data analysis is critical in chemistry for several reasons:

Validation of Hypotheses: Analyzing data helps scientists determine whether the results support or refute their initial hypotheses.

Identifying Trends and Patterns: Through data analysis, researchers can identify relationships between variables, recognize trends, and understand underlying chemical phenomena.

Guiding Future Research: Interpretation of data can inform future experiments, highlighting areas for further investigation or refinement of existing methodologies.

2. Types of Data in Chemistry

Data collected in a chemistry lab can be classified into two main categories:

Quantitative Data: This type involves numerical measurements and can be analyzed statistically. Examples include measurements of mass, volume, temperature, and concentration.

Qualitative Data: This type describes characteristics or properties that cannot be measured numerically. It includes observations such as color changes, precipitate formation, and odor detection.

3. Data Collection Techniques

Effective data collection is foundational for accurate analysis. Common techniques include:

Direct Measurements: Using laboratory equipment such as balances, graduated cylinders, and spectrophotometers to obtain precise numerical data.

Observational Methods: Documenting qualitative observations during experiments, such as changes in color or state.

Instrumental Methods: Utilizing advanced instruments (e.g., chromatographs, mass spectrometers) for collecting both qualitative and quantitative data.

4. Data Organization

Once data is collected, organizing it systematically is crucial for effective analysis. Common methods of organization include:

Tables: Presenting data in tabular form allows for easy comparison and clarity. Tables should include clear headings and units.

Graphs and Charts: Visual representations of data can facilitate the identification of trends and relationships. Common types include:

Bar Graphs: Useful for comparing discrete categories.

Line Graphs: Effective for showing trends over time or across a continuous range.

Scatter Plots: Useful for illustrating relationships between two quantitative variables.

5. Statistical Analysis

Statistical methods are vital for interpreting quantitative data accurately. Key concepts include:

Descriptive Statistics: Summarizing data using measures such as mean, median, mode, range, and standard deviation. These statistics provide insight into the central tendency and variability of the data.

Inferential Statistics: Techniques such as t-tests, ANOVA (Analysis of Variance), and regression analysis help researchers make inferences about populations based on sample data. These methods are crucial for determining the significance of results.

Error Analysis: Understanding and quantifying uncertainty is essential. Types of errors include systematic errors (consistent inaccuracies) and random errors (variability due to unpredictable factors). Researchers should calculate and report the margin of error and confidence intervals for their measurements.

6. Data Interpretation

Interpreting the results of data analysis involves drawing conclusions and making sense of the findings:

Comparison with Theoretical Values: Researchers should compare experimental results with theoretical predictions or literature values to assess accuracy.

Assessing Trends and Relationships: Analyzing graphs and statistical results helps identify correlations or causal relationships between variables. For example, the relationship between temperature and reaction rate may reveal insights into kinetic theory.

Identifying Anomalies: Any outliers or anomalies in the data should be critically examined. Researchers should consider whether these anomalies result from experimental error or if they indicate a new phenomenon worth investigating further.

7. Documentation and Reporting

Clear documentation and reporting of data analysis and interpretation are vital for transparency and reproducibility:

Lab Reports: A well-structured lab report should include sections such as introduction, methodology, results, discussion, and conclusion. Each section should be clear and concise, with appropriate data presented.

Citations and References: If the research involves previous studies or data, proper citation of sources is essential to maintain academic integrity.

Data Sharing: In many cases, sharing data with the scientific community can foster collaboration and enhance research validity. This may involve depositing data in public databases or publishing in peer-reviewed journals.

Data analysis and interpretation are essential skills in the chemistry laboratory. Through careful collection, organization, and analysis of data, researchers can validate their hypotheses, identify trends, and draw meaningful conclusions. A solid understanding of statistical methods and the ability to interpret results critically will empower chemists to contribute significantly to scientific knowledge and innovation. By adhering to best practices in data analysis, chemists can ensure their findings are robust, reproducible, and relevant, ultimately advancing the field of chemistry and its applications.

Chapter 14: Chemistry in Everyday Life

Household Chemistry: Cleaning Products, Food Additives, and Personal Care

Chemistry is an integral part of our daily lives, often in ways that go unnoticed. From the cleaning products we use in our homes to the food we consume and the personal care items we apply, chemistry plays a crucial role in ensuring safety, efficacy, and convenience. This chapter explores the chemical principles behind household products, their applications, and the implications for health and the environment.

1. Cleaning Products

Cleaning products are designed to remove dirt, stains, and contaminants from surfaces and materials. Their effectiveness is rooted in the chemistry of the ingredients used, which can be categorized into various types:

Surfactants: These are compounds that lower the surface tension of water, allowing it to spread and penetrate more effectively. Surfactants work by having both hydrophilic (water-attracting) and hydrophobic (water-repelling) properties, which help lift and wash away grease and grime. Common examples include sodium lauryl sulfate and linear alkylbenzene sulfonate.

Acids and Bases: Many cleaning agents contain acids or bases that react with stains or mineral deposits. For example, vinegar (acetic acid) is often used to remove hard water stains, while baking soda (sodium bicarbonate) acts as a mild abrasive and neutralizes odors. Stronger acids like hydrochloric acid are used in commercial cleaners for toilets and tough stains.

Enzymes: Some cleaning products contain enzymes that break down organic matter such as proteins, fats, and carbohydrates. Enzymatic cleaners are effective for laundry and dishwashing, as they help remove stains from food, sweat, and other organic materials.

Disinfectants: These products contain chemicals designed to kill bacteria, viruses, and fungi. Common disinfectants include bleach (sodium hypochlorite), hydrogen peroxide, and quaternary ammonium compounds. Understanding the chemistry of these disinfectants is essential for their effective use in maintaining hygiene.

2. Food Additives

Food additives are substances added to food to enhance its flavor, appearance, texture, or shelf-life. While some people may view additives with skepticism, many are essential for food preservation and safety. Common types of food additives include:

Preservatives: These chemicals prevent spoilage and extend the shelf life of food products. Examples include sodium benzoate, potassium sorbate, and nitrites, which inhibit the growth of bacteria and molds.

Coloring Agents: Artificial or natural colorants are added to enhance the visual appeal of food. Examples include tartrazine (Yellow 5) and beet juice powder. Understanding how these colorants interact with food can reveal insights into their stability and safety.

Flavor Enhancers: Substances like monosodium glutamate (MSG) enhance the umami flavor in foods. Flavor compounds may be derived from natural sources or synthesized chemically to provide desired tastes without increasing calories.

Emulsifiers: These additives stabilize mixtures of oil and water, preventing separation in products like salad dressings and mayonnaise. Lecithin, derived from soybeans, is a common emulsifier used in various foods.

Thickeners and Stabilizers: Ingredients such as xanthan gum, guar gum, and gelatin help achieve the desired texture and consistency in food products. These additives work by altering the viscosity of the mixture, which can enhance mouthfeel and appearance.

3. Personal Care Products

Personal care products encompass a wide range of items, including cosmetics, skincare, and hygiene products. The chemistry of these products is vital for their function, safety, and effectiveness. Key components include:

Emollients and Moisturizers: These substances hydrate the skin and improve its texture. Common ingredients include glycerin, hyaluronic acid, and various oils (e.g., coconut oil, shea butter). Understanding how these compounds interact with the skin helps formulate effective moisturizers.

Preservatives: Similar to food additives, personal care products require preservatives to prevent microbial growth. Common preservatives include parabens and phenoxyethanol. The safety and effectiveness of these preservatives are critical for ensuring product longevity and consumer safety.

Surfactants: Found in shampoos and cleansers, surfactants help lift dirt and oils from the skin and hair. Sodium lauryl sulfate and cocamidopropyl betaine are examples. The choice of surfactant affects the product's cleansing ability and gentleness.

Fragrances and Essential Oils: These are added to improve the sensory experience of personal care products. While natural essential oils provide pleasant aromas, synthetic fragrances may be designed to mimic specific scents. Understanding potential allergens in fragrances is important for consumer safety.

4. Safety and Environmental Considerations

While household chemistry provides many benefits, it also raises concerns regarding safety and environmental impact. Key considerations include:

Toxicity and Allergens: Some cleaning agents, food additives, and personal care products may cause allergic reactions or toxicity in certain individuals. Consumers should be educated about reading labels and understanding potential risks.

Environmental Impact: The production, use, and disposal of chemical products can have significant environmental consequences. For instance, phosphates in some detergents can contribute to water pollution, leading to algal blooms and ecosystem damage. Sustainable practices, such as using biodegradable products and reducing waste, are becoming increasingly important.

Regulatory Oversight: Many countries have regulatory bodies, such as the FDA (Food and Drug Administration) in the United States, that oversee the safety and efficacy of food additives and personal care products. Understanding the regulations can help consumers make informed choices.

Household chemistry plays a crucial role in everyday life, impacting our cleaning routines, food safety, and personal care practices. By understanding the chemical principles behind these products, consumers can make informed decisions that promote health and environmental sustainability. As science and technology continue to advance, the development of safer, more effective products will enhance our quality of life while minimizing negative impacts on health and the environment. Embracing the chemistry behind these everyday items empowers individuals to navigate their choices with greater awareness and responsibility.

Chemistry in Technology: Electronics, Materials, and Energy

The integration of chemistry into technology has transformed our lives, driving innovations in electronics, materials science, and energy production. This chapter explores the pivotal role chemistry plays in these sectors, highlighting how chemical principles underpin technological advancements that shape our modern world.

1. Chemistry in Electronics

The electronics industry relies heavily on chemistry to create components and materials that enhance performance, efficiency, and sustainability. Key areas where chemistry contributes to electronics include:

Semiconductors: Semiconductors are materials with electrical conductivity between that of conductors and insulators. Silicon, the most widely used semiconductor, is derived from silica (silicon dioxide) and undergoes various chemical processes to form pure silicon. Doping with elements like phosphorus or boron alters the electrical properties of silicon, enabling the production of p-type and n-type semiconductors, essential for transistors and diodes used in computers, smartphones, and other devices.

Conductive Materials: Conductive polymers, metals, and composites are vital for creating electrical connections in circuits. For example, copper is commonly used in wiring due to its excellent conductivity, while silver is used in specialized applications due to its superior conductive properties. Advances in nanotechnology have led to the development of nanowires and conductive inks that can be printed on flexible surfaces, expanding the possibilities for electronic devices.

Batteries and Energy Storage: The chemistry of batteries, including lithium-ion and lead-acid batteries, is crucial for energy storage in portable electronics and electric vehicles. Lithium-ion batteries, for example, rely on the electrochemical reactions between lithium ions and electrode materials during charging and discharging cycles. Research into new battery technologies, such as solid-state batteries and flow batteries, aims to improve energy density, safety, and sustainability.

Display Technologies: Chemistry is integral to developing advanced display technologies like organic light-emitting diodes (OLEDs) and liquid crystal displays (LCDs). OLEDs utilize organic compounds that emit light when an electric current is applied, allowing for thinner and more vibrant screens. LCDs employ liquid crystals that change alignment in response to electric fields, modulating light to create images.

2. Chemistry in Materials Science

Materials science combines principles of chemistry, physics, and engineering to develop new materials with tailored properties for various applications. Key areas include:

Polymers: Polymers are long-chain molecules that exhibit diverse properties, making them suitable for numerous applications. Chemistry plays a vital role in synthesizing polymers with specific characteristics, such as strength, flexibility, and thermal stability. Examples include polyethylene, used in packaging, and polycarbonate, known for its impact resistance in eyewear and electronics.

Nanomaterials: Nanotechnology focuses on materials at the nanoscale (1 to 100 nanometers), where unique properties arise due to size and surface area. Nanomaterials are used in drug delivery, sensors, and coatings. For instance, carbon nanotubes exhibit exceptional strength and conductivity, making them valuable in electronics and composite materials.

Smart Materials: Smart materials respond to environmental changes, such as temperature, pressure, or electric fields. Shape-memory alloys can return to a predetermined shape when heated, while piezoelectric materials generate electricity when mechanically stressed. These materials have applications in robotics, medical devices, and actuators.

Composite Materials: Composites combine two or more materials to achieve superior properties. For example, fiberglass, made from glass fibers embedded in a polymer matrix, is lightweight yet strong, making it ideal for construction, automotive, and aerospace applications. Chemistry is crucial in determining the interactions between components to optimize performance.

3. Chemistry in Energy

The energy sector is undergoing a significant transformation driven by chemistry, focusing on sustainable and efficient energy production, storage, and usage. Key areas include:

Renewable Energy: Chemistry is at the forefront of developing renewable energy sources, such as solar cells and biofuels. Photovoltaic cells convert sunlight into electricity through chemical reactions in semiconductor materials, while biofuels, derived from organic materials, provide alternative energy sources that can reduce dependence on fossil fuels.

Fuel Cells: Fuel cells generate electricity through electrochemical reactions between hydrogen and oxygen, producing water as a byproduct. Hydrogen fuel cells are gaining attention for their potential in transportation and stationary power applications due to their high efficiency and low emissions. The development of catalysts that enhance the reaction rates is a significant area of research in this field.

Energy Storage: Beyond traditional batteries, innovative energy storage solutions are being developed. Flow batteries, which use liquid electrolytes stored in separate tanks, allow for scalable energy storage. Supercapacitors, characterized by their ability to charge and discharge rapidly, are being explored for applications requiring quick bursts of energy, such as in hybrid vehicles.

Carbon Capture and Storage (CCS): As concerns about climate change grow, chemistry plays a crucial role in developing technologies for capturing and storing carbon dioxide emissions from industrial processes and power plants. This involves chemical reactions that either absorb CO2 or convert it into stable forms for storage or reuse, contributing to efforts to mitigate greenhouse gas emissions.

Chemistry is the backbone of technological advancement in electronics, materials science, and energy production. By understanding the chemical principles that govern these fields, researchers and engineers can continue to innovate and develop solutions that address global challenges, such as sustainability and resource efficiency. As technology evolves, the role of chemistry will remain central in shaping a future that leverages scientific knowledge to improve our lives and protect the environment. Embracing this intersection of chemistry and technology is essential for fostering innovation and addressing the complex issues facing society today.

Exploring Chemistry through Everyday Experiences

Chemistry is not just confined to laboratories; it permeates our everyday experiences and shapes the world around us. This chapter explores how chemistry influences daily life through various interactions, products, and processes, illustrating its relevance and importance. By examining common occurrences, substances, and phenomena, we can appreciate the profound impact of chemistry on our health, environment, and technology.

1. Chemistry in Food and Cooking

Food chemistry is a fascinating domain that examines how the chemical composition of food affects flavor, texture, and nutritional value. Here are some key aspects:

Cooking Processes: Many cooking techniques, such as baking, boiling, and frying, involve chemical reactions. For example, the Maillard reaction occurs when proteins and sugars in food undergo complex reactions during cooking, leading to browning and the development of new flavors. Similarly, caramelization transforms sugars into rich, brown flavors through a series of chemical reactions.

Preservation: Chemistry plays a crucial role in food preservation methods. Techniques like canning, pickling, and fermenting rely on chemical reactions to inhibit microbial growth and prolong shelf life. For instance, adding salt to vegetables in the pickling process creates an environment that prevents spoilage by harmful bacteria.

Nutritional Chemistry: Understanding the chemical composition of food allows us to assess its nutritional value. Nutrients such as carbohydrates, proteins, fats, vitamins, and minerals have specific chemical structures that determine how they function in our bodies. This knowledge helps us make informed dietary choices and understand the health implications of various foods.

2. Chemistry in Personal Care Products

From soaps and shampoos to lotions and cosmetics, chemistry is integral to the formulation and efficacy of personal care products:

Surfactants in Cleaning: Surfactants are chemical compounds that lower the surface tension of liquids, allowing them to spread and penetrate more easily. In soaps and detergents, surfactants facilitate the removal of dirt and grease by emulsifying oils, making them soluble in water. This process is crucial for effective cleaning.

Emulsions in Lotions and Creams: Many personal care products, such as moisturizers, are emulsions—mixtures of oil and water stabilized by emulsifying agents. Understanding the chemistry behind emulsions allows formulators to create products with the desired consistency and stability.

Fragrance Chemistry: The scent of personal care products is often derived from a complex mixture of aromatic compounds. Chemists carefully select and combine these compounds to create appealing fragrances, understanding how different chemicals interact with each other and our olfactory senses.

Cosmetic Chemistry: The formulation of makeup involves a variety of chemicals, including pigments, preservatives, and stabilizers. Understanding the chemistry behind these ingredients ensures safety and efficacy, allowing for the creation of long-lasting and skin-friendly products.

3. Chemistry in Cleaning Products

Household cleaning products are a prime example of applied chemistry, designed to enhance cleanliness and hygiene:

Acids and Bases: Many cleaning products utilize acidic or basic compounds to effectively remove stains and disinfect surfaces. For instance, vinegar (acetic acid) is effective in removing mineral deposits, while baking soda (sodium bicarbonate) acts as a mild abrasive and neutralizes odors. Understanding the pH of cleaning products can help consumers choose the right product for specific tasks.

Disinfectants: The effectiveness of disinfectants relies on their chemical properties. For example, bleach (sodium hypochlorite) kills bacteria and viruses by breaking down their cellular structure. Understanding the mechanism of action of various disinfectants is crucial for effective cleaning and sanitization.

Enzymatic Cleaners: Enzymatic cleaners contain specific enzymes that break down organic matter, such as proteins and fats. These products are particularly effective for removing stains from carpets and fabrics. Knowing how enzymes function at a molecular level can help consumers choose the right cleaning solution.

4. Chemistry in Health and Medicine

Chemistry is fundamental to medicine, influencing drug development, diagnostics, and treatment:

Pharmaceutical Chemistry: The development of medications involves understanding the chemical properties of compounds and how they interact with biological systems. Pharmacists use chemistry to formulate drugs, ensuring they are effective, safe, and bioavailable. Knowledge of chemical reactions helps in designing new drugs that target specific diseases or conditions.

Diagnostics: Chemistry underpins various diagnostic techniques, such as blood tests and imaging technologies. For instance, the use of contrast agents in medical imaging relies on their chemical properties to enhance visibility in X-rays or MRIs.

Vaccines: Vaccines are developed using complex biochemical processes that stimulate the immune system. Understanding the chemistry of antigens and adjuvants is crucial for creating effective vaccines that provide immunity against diseases.

5. Chemistry in Environmental Issues

Chemistry helps us understand environmental challenges and develop sustainable solutions:

Pollution Chemistry: The study of pollutants—substances that contaminate the environment—relies on chemical analysis to understand their sources, behavior, and effects. For example, understanding the chemical reactions that occur during combustion helps in developing cleaner fuels and reducing air pollution.

Green Chemistry: This innovative approach aims to design chemical processes that minimize waste and reduce harmful effects on the environment. Green chemistry principles encourage the use of renewable resources, non-toxic solvents, and energy-efficient methods, promoting sustainability in chemical production.

Water Treatment: Chemistry is essential for water purification and treatment processes. Understanding the chemical reactions involved in removing contaminants, such as chlorine disinfection or activated carbon filtration, helps ensure safe drinking water.

6. Chemistry in Technology and Innovation

Advancements in technology are often driven by chemical innovations, impacting various industries:

Materials Science: The development of new materials, such as polymers, nanomaterials, and composites, relies on chemical knowledge. These materials find applications in electronics, construction, and manufacturing, enhancing product performance and efficiency.

Sustainable Energy: Chemistry plays a crucial role in the development of renewable energy sources, such as solar cells and biofuels. Understanding the chemical processes involved in energy conversion and storage is vital for creating sustainable solutions to meet growing energy demands.

Exploring chemistry through everyday experiences reveals its profound influence on our lives. From the food we eat and the products we use to the medicines we rely on, chemistry is integral to understanding and improving our world. By recognizing the chemical principles at play in our daily lives, we can make informed choices that promote health, sustainability, and technological advancement. As we continue to explore the intersection of chemistry and everyday life, we gain a deeper appreciation for the science that shapes our experiences and drives innovation.

Chapter 15: Advanced Topics (Optional)

Introduction to Quantum Chemistry

Quantum chemistry is a fascinating branch of chemistry that applies the principles of quantum mechanics to understand and predict the behavior of matter at the atomic and molecular levels. Unlike classical chemistry, which often relies on macroscopic observations, quantum chemistry delves into the subatomic realm to explain the properties and interactions of particles such as atoms, electrons, and molecules. This chapter provides an overview of quantum chemistry, its fundamental concepts, key principles, and applications in various fields.

1. Historical Background

The development of quantum chemistry emerged in the early 20th century, coinciding with significant advancements in physics and the understanding of atomic structure. The transition from classical mechanics to quantum mechanics was driven by the limitations of classical theories to explain phenomena such as atomic spectra and the behavior of electrons.

Max Planck and Quantum Theory: In 1900, Max Planck proposed that energy is quantized, introducing the concept of "quanta" to explain black-body radiation. This idea laid the groundwork for later developments in quantum theory.

Niels Bohr and the Bohr Model: Niels Bohr's model of the hydrogen atom (1913) incorporated quantization by suggesting that electrons occupy discrete energy levels. This model successfully explained the spectral lines of hydrogen and marked a significant shift in understanding atomic structure.

Wave-Particle Duality: The concept of wave-particle duality, articulated by Louis de Broglie, posited that particles such as electrons exhibit both wave-like and particle-like properties. This duality is foundational in quantum mechanics and has profound implications for the study of atomic and molecular behavior.

Schrödinger Equation: In 1926, Erwin Schrödinger formulated a mathematical equation describing how the quantum state of a physical system changes over time. The Schrödinger equation is central to quantum chemistry, as it allows for the calculation of wave functions and energy levels of quantum systems.

2. Fundamental Concepts of Quantum Chemistry

Several key concepts form the foundation of quantum chemistry, enabling a deeper understanding of atomic and molecular behavior.

Wave Functions and Quantum States: A wave function, represented by the symbol Ψ (psi), is a mathematical function that describes the quantum state of a system. The square of the wave function's absolute value, $|\Psi|^2$, gives the probability density of finding a particle in a particular location. Wave functions provide critical information about the energy levels and spatial distributions of electrons in atoms and molecules.

Quantization of Energy Levels: In quantum chemistry, energy levels are quantized, meaning that electrons can only occupy specific energy states. This principle explains why atoms emit or absorb light at distinct wavelengths, resulting in the characteristic spectra observed for different elements.

Uncertainty Principle: Formulated by Werner Heisenberg, the uncertainty principle states that it is impossible to simultaneously know both the position and momentum of a particle with absolute certainty. This inherent uncertainty challenges classical intuitions about particle behavior and leads to a probabilistic interpretation of quantum states.

Quantum Superposition: Quantum superposition is the principle that a quantum system can exist in multiple states simultaneously until it is measured. This concept is pivotal in understanding phenomena such as electron configuration and the behavior of molecules in chemical reactions.

3. Molecular Orbital Theory

Molecular orbital (MO) theory is a significant application of quantum chemistry that describes the electronic structure of molecules. Unlike valence bond theory, which focuses on individual bonds, MO theory considers the combination of atomic orbitals to form molecular orbitals that are spread over the entire molecule.

Formation of Molecular Orbitals: When atomic orbitals combine, they can form bonding and antibonding molecular orbitals. Bonding orbitals are lower in energy and promote stability, while antibonding orbitals are higher in energy and destabilize the molecule. The distribution of electrons in these orbitals determines molecular stability and reactivity.

Electron Configuration in Molecules: MO theory allows chemists to determine the electron configuration of molecules, providing insights into their chemical properties, bonding behavior, and magnetic characteristics. For example, the presence of unpaired electrons in molecular orbitals can indicate whether a molecule is paramagnetic (attracted to magnetic fields) or diamagnetic (repelled by magnetic fields).

Delocalization: MO theory also explains the concept of delocalization, where electrons are not confined to a single bond or atom but are spread over several atoms in a molecule. This phenomenon is crucial in understanding resonance structures and the stability of conjugated systems.

4. Applications of Quantum Chemistry

Quantum chemistry has numerous applications across various fields, including materials science, biochemistry, and pharmaceuticals.

Drug Design: Quantum chemistry plays a vital role in computational drug design, where molecular modeling and simulations help predict the interactions between drug candidates and biological targets. By understanding the electronic structure of molecules, researchers can optimize drug efficacy and minimize side effects.

Material Science: Quantum chemistry aids in the design and characterization of new materials with specific properties. For instance, researchers can use quantum mechanical calculations to understand the electronic properties of semiconductors, conductors, and insulators, paving the way for advancements in nanotechnology and electronics.

Spectroscopy: Quantum chemistry provides a theoretical framework for interpreting spectroscopic data, enabling scientists to study molecular structures and dynamics. Techniques such as infrared (IR) spectroscopy and nuclear magnetic resonance (NMR) spectroscopy rely on quantum principles to probe molecular behavior.

Catalysis: Understanding the quantum mechanics of catalysts can lead to the development of more efficient catalytic processes. Quantum chemistry helps identify the active sites and reaction pathways in catalytic reactions, enhancing the design of catalysts for industrial applications.

5. Challenges and Future Directions

Despite its successes, quantum chemistry faces several challenges and limitations:

Computational Complexity: Quantum mechanical calculations can be computationally intensive, especially for large molecules or complex systems. The development of more efficient algorithms and computational methods is essential for expanding the applicability of quantum chemistry.

Interpretation of Results: The probabilistic nature of quantum mechanics can lead to challenges in interpreting results and drawing definitive conclusions. Ongoing research seeks to develop clearer frameworks for understanding quantum systems and their behaviors.

Integration with Classical Chemistry: Bridging the gap between quantum chemistry and classical chemistry remains an area of active research. Understanding how quantum effects influence macroscopic properties and behaviors is crucial for a comprehensive understanding of chemical systems.

Quantum chemistry represents a profound intersection of chemistry and physics, providing invaluable insights into the behavior of matter at the molecular level. By applying the principles of quantum mechanics, chemists can predict and understand the properties of atoms and molecules, leading to significant advancements in various fields. As computational power continues to grow and new theoretical developments emerge, the future of quantum chemistry holds exciting potential for unraveling the complexities of the chemical world and driving innovation in science and technology.

Chemical Bonding Theories

Chemical bonding theories are essential for understanding how atoms combine to form molecules. These theories explain the nature of chemical bonds, the arrangement of atoms in molecules, and the physical and chemical properties resulting from these arrangements. This chapter will delve into several key chemical bonding theories, including the Valence Bond Theory (VBT), Molecular Orbital Theory (MOT), and the concept of hybridization, while highlighting their applications and limitations.

1. Valence Bond Theory (VBT)

Valence Bond Theory is one of the earliest and most widely taught theories of chemical bonding. Developed in the 1920s and 1930s by Linus Pauling and others, VBT explains how atoms share electrons to form covalent bonds.

A. Basic Principles

Orbital Overlap: According to VBT, a covalent bond forms when atomic orbitals from two atoms overlap. This overlap allows for the sharing of electrons between the two nuclei, stabilizing the bond. The strength of the bond increases with greater overlap of the involved orbitals.

Types of Bonds: VBT distinguishes between sigma (σ) and pi (π) bonds:

Sigma Bonds: Formed by the head-on overlap of atomic orbitals. This can occur between two s orbitals, an s and a p orbital, or two p orbitals. Sigma bonds are characterized by their cylindrical symmetry around the bond axis.

Pi Bonds: Formed by the side-to-side overlap of p orbitals. Pi bonds are present in double and triple bonds, alongside sigma bonds. They restrict rotation around the bond axis due to their orbital orientation.

B. Hybridization

Hybridization is a concept within VBT that explains the geometry of molecular structures. It describes the mixing of atomic orbitals to form new, equivalent hybrid orbitals that can overlap with other atomic orbitals.

Types of Hybridization:

sp Hybridization: Involves the mixing of one s orbital and one p orbital, forming two equivalent sp hybrid orbitals arranged linearly (180° apart). This hybridization is typical for molecules like $BeCl_2$ and acetylene (C_2H_2).

sp² Hybridization: Involves the mixing of one s orbital and two p orbitals to create three equivalent sp^2 hybrid orbitals arranged trigonal planar (120° apart). This is seen in molecules like BF_3 and ethylene (C_2H_4).

sp³ Hybridization: Involves one s orbital and three p orbitals, producing four equivalent sp^3 hybrid orbitals arranged tetrahedrally (109.5° apart). This is characteristic of molecules like methane (CH_4).

Other Hybridizations: Complex molecules may involve higher hybridizations, such as sp^3d (for trigonal bipyramidal structures) and sp^3d^2 (for octahedral structures).

C. Limitations of Valence Bond Theory

While VBT provides a useful framework for understanding molecular bonding, it has limitations:

Inability to Explain Magnetic Properties: VBT cannot account for the paramagnetism observed in certain molecules, such as O_2, which contain unpaired electrons.

Resonance Structures: VBT struggles to explain resonance, where a molecule can be represented by multiple valid Lewis structures. This limitation makes it difficult to accurately depict molecules like benzene (C_6H_6).

Quantitative Predictions: VBT is often qualitative and does not provide precise predictions of molecular properties.

2. Molecular Orbital Theory (MOT)

Molecular Orbital Theory provides a more sophisticated approach to understanding chemical bonding. Developed in the early 20th century, MOT treats electrons in a molecule as occupying molecular orbitals that are delocalized over the entire molecule rather than being associated with individual bonds.

A. Basic Principles

Formation of Molecular Orbitals: When atomic orbitals combine, they form molecular orbitals that can be classified as bonding or antibonding:

Bonding Orbitals: Formed by the constructive interference of atomic orbitals, resulting in a lower energy state and increased stability.

Antibonding Orbitals: Formed by the destructive interference of atomic orbitals, resulting in a higher energy state and decreased stability.

Electron Configuration in Molecules: Electrons fill molecular orbitals according to the same principles as atomic orbitals, with the lowest energy orbitals filling first. The filling order and arrangement of electrons in these orbitals determine the molecule's properties and reactivity.

B. Molecular Orbital Diagrams

Molecular orbital diagrams visually represent the energy levels of molecular orbitals and the distribution of electrons in those orbitals. These diagrams aid in understanding the bonding and antibonding interactions within a molecule.

H_2 Molecule Example: In hydrogen (H_2), the two 1s atomic orbitals combine to form one bonding σ(1s) orbital and one antibonding σ*(1s) orbital. The two electrons fill the bonding orbital, resulting in a stable H_2 molecule.

Homodiatomic Molecules: For molecules like O_2 and N_2, molecular orbital diagrams help visualize the relative energies of bonding and antibonding orbitals, providing insights into magnetic properties and bond order.

C. Applications of Molecular Orbital Theory

MOT is particularly useful in understanding:

Magnetic Properties: MOT explains why certain molecules, such as O_2, are paramagnetic due to unpaired electrons in degenerate molecular orbitals.

Bonding in Complex Molecules: MOT can be applied to large and complex molecules, providing a more comprehensive understanding of bonding in transition metal complexes and coordination compounds.

D. Limitations of Molecular Orbital Theory

Despite its strengths, MOT also has limitations:

Complex Calculations: The mathematical complexity of molecular orbital calculations can be daunting, particularly for large molecules or systems with multiple interacting electrons.

Less Intuitive: For beginners, MOT can be less intuitive compared to VBT, which relies on visualizing orbital overlaps and hybridization.

3. Comparison of Valence Bond Theory and Molecular Orbital Theory

Both VBT and MOT are valuable in understanding chemical bonding, but they approach the subject from different perspectives:

VBT focuses on localized electron pairs and overlaps of atomic orbitals, making it particularly useful for simpler molecules and understanding hybridization.

MOT provides a delocalized view of electrons and can account for phenomena that VBT cannot, such as paramagnetism and resonance.

Chemical bonding theories, including Valence Bond Theory and Molecular Orbital Theory, are essential for understanding the nature of chemical bonds and the behavior of molecules. While VBT offers insights into localized bonding and hybridization, MOT expands the understanding of electron delocalization and molecular properties. Together, these theories provide a comprehensive framework for chemists to explain, predict, and manipulate the chemical behavior of substances in various fields, from materials science to drug design. As research continues to advance in quantum chemistry and computational methods, these bonding theories will evolve, leading to new discoveries and applications in chemistry and beyond.

Spectroscopy and Instrumental Methods

Spectroscopy and instrumental methods are pivotal in the field of chemistry, allowing scientists to analyze the structure, composition, and properties of substances. This chapter will explore the fundamental principles of various spectroscopic techniques and instrumental methods, their applications in chemical analysis, and the theoretical foundations that underpin these powerful tools.

1. Introduction to Spectroscopy

Spectroscopy is the study of the interaction between matter and electromagnetic radiation. It encompasses a range of techniques used to identify and quantify substances based on how they absorb, emit, or scatter light. The electromagnetic spectrum includes various types of radiation, from gamma rays to radio waves, with each type providing unique insights into the material being analyzed.

A. Basic Principles of Spectroscopy

Electromagnetic Radiation: At the core of spectroscopy is electromagnetic radiation, which travels in waves characterized by wavelength and frequency. The interaction of this radiation with matter leads to various phenomena, such as absorption, emission, and scattering.

Quantized Energy Levels: In atoms and molecules, electrons exist at specific energy levels. When energy is absorbed or emitted, transitions occur between these quantized levels, producing characteristic spectra.

Spectral Lines: The emitted or absorbed light appears as spectral lines, which can be used to identify the elemental or molecular composition of a sample. The position, intensity, and shape of these lines provide valuable information about the electronic structure of the substance.

2. Types of Spectroscopy

Numerous spectroscopic techniques are employed in chemistry, each utilizing different regions of the electromagnetic spectrum and varying principles of interaction. Below are some of the most commonly used types:

A. Infrared (IR) Spectroscopy

Principle: Infrared spectroscopy measures the absorption of infrared radiation by a sample, which causes molecular vibrations. Different functional groups absorb IR radiation at characteristic frequencies.

Applications: IR spectroscopy is widely used to identify functional groups in organic compounds, determine molecular structure, and analyze mixtures. It is invaluable in fields such as pharmaceuticals, materials science, and environmental chemistry.

FTIR (Fourier Transform Infrared Spectroscopy): A modern advancement, FTIR collects data rapidly and provides high-resolution spectra by transforming the raw data into a usable spectrum using mathematical algorithms.

B. Nuclear Magnetic Resonance (NMR) Spectroscopy

Principle: NMR spectroscopy relies on the magnetic properties of certain nuclei, primarily hydrogen (^{1}H) and carbon (^{13}C). When placed in a strong magnetic field and exposed to radiofrequency radiation, nuclei can absorb energy and transition between spin states.

Applications: NMR is essential for determining the structure of organic compounds, elucidating molecular dynamics, and analyzing complex mixtures. It provides detailed information about the environment of specific nuclei within molecules.

2D and 3D NMR Techniques: Advanced NMR techniques, such as two-dimensional (2D) and three-dimensional (3D) NMR, allow for even greater structural insights by correlating signals from different nuclei.

C. Ultraviolet-Visible (UV-Vis) Spectroscopy

Principle: UV-Vis spectroscopy measures the absorption of ultraviolet and visible light by a sample, which excites electrons to higher energy levels. The resulting spectrum indicates the wavelengths of light absorbed.

Applications: This technique is commonly used to analyze the concentration of colored solutions, study electronic transitions in molecules, and monitor chemical reactions. It is widely employed in fields like biochemistry, environmental science, and quality control in pharmaceuticals.

D. Mass Spectrometry (MS)

Principle: Mass spectrometry measures the mass-to-charge ratio of ions generated from a sample. The sample is ionized, and the resulting ions are separated based on their mass and charge using electric and magnetic fields.

Applications: MS is a powerful tool for identifying and quantifying compounds, determining molecular weights, and analyzing complex mixtures. It is extensively used in proteomics, metabolomics, and environmental analysis.

Coupled Techniques: MS is often coupled with chromatographic techniques (e.g., GC-MS, LC-MS) to separate mixtures before mass analysis, enhancing the technique's sensitivity and resolution.

E. Raman Spectroscopy

Principle: Raman spectroscopy is based on the inelastic scattering of monochromatic light (usually from a laser) by molecules. The energy change of scattered light provides information about molecular vibrations and chemical bonds.

Applications: Raman spectroscopy is used to analyze molecular structure, identify chemical compounds, and study phase transitions. It is particularly useful for studying solids, liquids, and gases without extensive sample preparation.

3. Instrumental Methods

Instrumental methods refer to a broad range of analytical techniques that utilize sophisticated instruments for chemical analysis. These methods often incorporate various spectroscopic techniques and provide quantitative and qualitative information about samples.

A. Chromatography

Chromatography is a technique used to separate components of a mixture based on their interactions with a stationary phase and a mobile phase. Common types of chromatography include:

Gas Chromatography (GC): Utilizes a gas as the mobile phase to separate volatile compounds. Ideal for analyzing gases and volatile liquids.

Liquid Chromatography (LC): Involves a liquid mobile phase, suitable for a wide range of samples, including complex biological mixtures.

Chromatography is often coupled with MS (GC-MS, LC-MS) for enhanced sensitivity and separation capabilities.

B. Electrochemical Methods

Electrochemical methods involve the measurement of electrical properties (e.g., current, voltage) in chemical reactions. Techniques include:

Voltammetry: Measures current as a function of applied voltage, providing information about the electrochemical behavior of analytes.

Potentiometry: Involves measuring the voltage of electrochemical cells to determine the concentration of ions in solution (e.g., pH measurement).

Electrochemical methods are widely used in environmental monitoring, clinical diagnostics, and industrial applications.

4. Data Analysis in Spectroscopy

The analysis of spectral data is crucial for interpreting results and extracting meaningful information. Key steps in data analysis include:

Peak Identification: Recognizing peaks in spectra corresponds to specific energy transitions or molecular vibrations.

Quantification: Determining the concentration of analytes based on the intensity of spectral peaks, often using calibration curves.

Comparison with Standards: Comparing obtained spectra with known standards to identify unknown compounds and confirm the presence of specific functional groups.

Spectroscopy and instrumental methods are indispensable tools in modern chemistry, enabling scientists to explore the structure and behavior of molecules at an atomic level. From understanding molecular interactions to analyzing complex mixtures, these techniques provide crucial insights across various fields, including pharmaceuticals, environmental science, and materials development. As technology continues to advance, the capabilities of spectroscopic and instrumental methods will expand, facilitating even more sophisticated analyses and driving innovation in scientific research and industry.

Nanotechnology in Chemistry

Nanotechnology, a field that deals with the manipulation and understanding of materials at the nanometer scale (1 to 100 nanometers), is revolutionizing various branches of chemistry. This chapter explores the principles of nanotechnology, its applications, and the impact it has on the field of chemistry and beyond.

1. Introduction to Nanotechnology

Nanotechnology refers to the engineering and manipulation of matter at the nanoscale, where unique physical and chemical properties emerge due to size and surface effects. At this scale, materials often exhibit distinct behaviors that differ from their bulk counterparts, making nanotechnology a multidisciplinary field that encompasses chemistry, physics, materials science, and biology.

A. Nanoscale Definition and Significance

Definition: A nanometer (nm) is one-billionth of a meter, which is about 100,000 times smaller than the diameter of a human hair. At this scale, the properties of materials can change dramatically.

Significance: The properties of materials can differ significantly at the nanoscale due to increased surface area-to-volume ratios, quantum effects, and other factors. For example, gold nanoparticles exhibit different optical properties than bulk gold, making them useful in various applications.

2. Synthesis of Nanomaterials

The creation of nanomaterials involves several techniques that can be broadly categorized into two approaches: top-down and bottom-up.

A. Top-Down Approach

This approach involves breaking down larger materials into nanoscale particles. Common methods include:

Mechanical Milling: Grinding bulk materials into fine nanoparticles.

Lithography: Utilizing patterns to create nanoscale structures on surfaces.

B. Bottom-Up Approach

In this approach, nanoscale materials are built from smaller units, such as atoms or molecules. Methods include:

Chemical Vapor Deposition (CVD): A chemical process used to produce thin films of nanomaterials.

Sol-Gel Processes: Transforming a solution into a solid gel phase, which can be further processed to create nanoparticles.

C. Biological Methods

Biological systems can also be harnessed to synthesize nanomaterials, often referred to as "green nanotechnology." Microorganisms and plants can produce nanoparticles through natural metabolic processes, offering an environmentally friendly alternative to traditional synthesis methods.

3. Characterization of Nanomaterials

Characterizing nanomaterials is crucial to understanding their properties and potential applications. Various techniques are employed:

A. Electron Microscopy

Transmission Electron Microscopy (TEM): Provides high-resolution images of nanoparticles, allowing for the study of their structure and morphology.

Scanning Electron Microscopy (SEM): Offers three-dimensional images of the surface morphology of nanomaterials.

B. Atomic Force Microscopy (AFM)

AFM allows for the imaging of surfaces at the nanoscale by scanning a sharp tip over the sample surface, providing information about surface topology and mechanical properties.

C. X-ray Diffraction (XRD)

XRD is used to determine the crystalline structure and phase identification of nanomaterials, providing insights into their physical properties.

4. Applications of Nanotechnology in Chemistry

Nanotechnology has numerous applications across various fields, with significant implications for chemistry. Here are some key areas of impact:

A. Drug Delivery Systems

Nanotechnology has transformed the field of pharmaceuticals by enabling targeted drug delivery. Nanoparticles can encapsulate drugs, enhancing solubility and stability while allowing for controlled release. Key advantages include:

Targeted Delivery: Nanoparticles can be engineered to target specific tissues or cells, minimizing side effects and improving therapeutic efficacy.

Enhanced Bioavailability: Nanoparticles can increase the absorption and distribution of drugs within the body.

B. Catalysis

Nanocatalysts are increasingly utilized to enhance reaction rates and selectivity in chemical processes. Benefits include:

Increased Surface Area: The high surface area of nanoparticles increases the availability of active sites for reactions.

Improved Reaction Efficiency: Nanocatalysts can often operate at lower temperatures and pressures, making chemical processes more energy-efficient.

C. Environmental Remediation

Nanotechnology plays a crucial role in addressing environmental challenges, such as:

Pollution Control: Nanomaterials can be employed to remove contaminants from water and air. For instance, nanoscale zero-valent iron is used to treat groundwater contaminated with heavy metals and organic pollutants.

Sensing and Detection: Nanosensors can detect environmental pollutants at extremely low concentrations, providing real-time monitoring of environmental conditions.

D. Energy Storage and Conversion

Nanotechnology is advancing the development of more efficient energy systems, including:

Nanostructured Batteries: Enhancing the capacity and lifespan of batteries through the use of nanomaterials, such as carbon nanotubes and metal oxides.

Photovoltaics: Nanotechnology enables the creation of more efficient solar cells, improving light absorption and energy conversion efficiency.

5. Challenges and Future Directions

While nanotechnology holds immense promise, several challenges must be addressed:

A. Safety and Toxicity

The potential toxicity of nanomaterials is a critical concern. Research is ongoing to understand the health and environmental impacts of nanomaterials, and regulations are needed to ensure safe handling and disposal.

B. Scalability and Production

Developing scalable production methods for nanomaterials that are cost-effective and environmentally friendly is essential for widespread adoption in industrial applications.

C. Ethical Considerations

As with any emerging technology, ethical considerations regarding the use of nanotechnology, including privacy issues related to nanosensors and potential misuse, must be addressed.

6. Conclusion

Nanotechnology represents a frontier in chemistry, offering transformative solutions across various fields, from medicine to environmental science. By enabling the manipulation of materials at the atomic and molecular levels, nanotechnology opens new avenues for research and innovation, promising advancements that can significantly impact society. As this field continues to evolve, ongoing research and collaboration among scientists, engineers, and policymakers will be vital in addressing the challenges and harnessing the potential of nanotechnology for the greater good.

Chapter 16: Conclusion

Common Chemical Formulas and Equations

In this concluding chapter, we summarize key concepts from the study of chemistry and provide an overview of common chemical formulas and equations that are essential for understanding chemical reactions and processes. This chapter serves as a valuable reference for students and practitioners alike, highlighting the significance of chemical formulas in the broader context of chemistry.

1. Understanding Chemical Formulas

Chemical formulas are symbolic representations of substances that indicate the types and numbers of atoms present in a molecule or compound. They play a crucial role in conveying information about the composition and structure of chemical substances. Here, we delve into the various types of chemical formulas and their significance:

A. Empirical Formula

The empirical formula of a compound represents the simplest whole-number ratio of atoms of each element in the compound. It does not provide information about the actual number of atoms in a molecule but rather the relative proportions. For example, the empirical formula for hydrogen peroxide (H_2O_2) is HO, indicating a 1:1 ratio of hydrogen to oxygen.

B. Molecular Formula

The molecular formula provides the actual number of atoms of each element in a molecule of the compound. It is derived from the empirical formula and is crucial for understanding the specific composition of a substance. For instance, the molecular formula for glucose is $C_6H_{12}O_6$, indicating that each molecule contains six carbon atoms, twelve hydrogen atoms, and six oxygen atoms.

C. Structural Formula

The structural formula offers a visual representation of the arrangement of atoms within a molecule. It illustrates how atoms are bonded together, including the types of bonds (single, double, or triple) and the connectivity of atoms. For example, the structural formula of ethanol (C_2H_5OH) shows the arrangement of carbon, hydrogen, and oxygen atoms.

2. Common Chemical Equations

Chemical equations represent chemical reactions, illustrating the transformation of reactants into products. They are essential for understanding the stoichiometry of reactions and predicting the outcomes of chemical processes. Here, we outline some common types of chemical equations:

A. Combination Reactions

In combination reactions, two or more substances combine to form a single product. The general form is:

$$A + B \rightarrow AB$$

Example: The formation of water from hydrogen and oxygen:

$$2H_2 + O_2 \rightarrow 2H_2O$$

B. Decomposition Reactions

Decomposition reactions involve the breakdown of a single compound into two or more simpler substances. The general form is:

$$AB \rightarrow A + B$$

Example: The decomposition of potassium chlorate into potassium chloride and oxygen:

$$2KClO_3 \rightarrow 2KCl + 3O_2$$

C. Single Replacement Reactions

In single replacement reactions, one element replaces another in a compound. The general form is:

$$A + BC \rightarrow AC + B$$

Example: The reaction of zinc with hydrochloric acid:

$$Zn + 2HCl \rightarrow ZnCl_2 + H_2$$

D. Double Replacement Reactions

In double replacement reactions, the anions and cations of two different compounds exchange places to form two new compounds. The general form is:

$$AB + CD \rightarrow AD + CB$$

Example: The reaction between silver nitrate and sodium chloride:

$$AgNO_3 + NaCl \rightarrow AgCl + NaNO_3$$

E. Combustion Reactions

Combustion reactions involve the reaction of a substance with oxygen, producing heat and light. Commonly, hydrocarbon combustion produces carbon dioxide and water. The general form is:

$$C_xH_y + O_2 \rightarrow CO_2 + H_2O$$

Example: The combustion of methane:

$$CH_4 + 2O_2 \rightarrow CO_2 + 2H_2O$$

3. Balancing Chemical Equations

One of the fundamental principles of chemistry is the law of conservation of mass, which states that matter cannot be created or destroyed in a chemical reaction. Therefore, balancing chemical equations is essential to ensure that the number of atoms of each element is the same on both sides of the equation.

A. Steps to Balance Chemical Equations

Write the unbalanced equation: Start with the correct formulas for reactants and products.

Count the atoms of each element: List the number of atoms for each element on both sides of the equation.

Adjust coefficients: Change the coefficients in front of the formulas to balance the number of atoms for each element.

Recheck: Ensure that the number of atoms is equal on both sides and that the coefficients are in the simplest whole-number ratio.

B. Example of Balancing a Reaction

Balancing the combustion of propane (C_3H_8):

$$C_3H_8 + O_2 \rightarrow CO_2 + H_2O$$

Count the atoms:

Reactants: C=3, H=8, O=2

Products: C=1, H=2, O=3 (1 CO_2 and 1 H_2O).

Balance carbon by adding coefficients:

$$C_3H_8 + O_2 \rightarrow 3CO_2 + H_2O$$

Balance hydrogen:

$$C_3H_8 + O_2 \rightarrow 3CO_2 + 4H_2O$$

Balance oxygen:

$$C_3H_8 + 5O_2 \rightarrow 3CO_2 + 4H_2O$$

The balanced equation becomes:

$$C_3H_8 + 5O_2 \rightarrow 3CO_2 + 4H_2O$$

Understanding common chemical formulas and equations is essential for grasping the principles of chemistry. This knowledge forms the basis for more advanced studies and applications in the field. By mastering the art of writing and balancing chemical equations, students gain a deeper insight into the behavior of substances, their interactions, and the transformations that occur during chemical reactions.

As you continue your journey in chemistry, refer back to these common formulas and equations, as they will serve as fundamental tools in your scientific endeavors. Whether in academic studies or practical applications, the ability to understand and manipulate chemical formulas is crucial for success in the world of chemistry.

The Periodic Table of Elements

The periodic table of elements is one of the most significant achievements in the history of chemistry, serving as an essential framework for understanding the relationships between different elements. This chapter concludes the exploration of chemistry by summarizing the importance of the periodic table, its organization, and its implications in various fields.

1. Historical Background

The development of the periodic table dates back to the 19th century when scientists sought to classify elements based on their properties and atomic masses. The contributions of key figures such as Dmitri Mendeleev, who created the first widely recognized periodic table in 1869, laid the groundwork for its modern form. Mendeleev's arrangement of elements revealed patterns in their properties and allowed for the prediction of undiscovered elements.

Mendeleev organized elements in rows based on increasing atomic mass, placing elements with similar chemical properties in columns. His table was revolutionary, as it highlighted periodicity—the tendency of elements to exhibit similar properties at regular intervals when arranged by atomic mass.

2. Structure of the Periodic Table

The modern periodic table is organized by increasing atomic number, which is the number of protons in an atom's nucleus. The arrangement of elements into periods (horizontal rows) and groups (vertical columns) reflects the periodic trends in chemical and physical properties.

A. Periods

The periodic table has seven periods, each corresponding to the filling of electron shells around an atom's nucleus. As you move from left to right across a period, the atomic number increases, and elements display varying properties:

Period 1: Contains two elements, hydrogen (H) and helium (He).

Period 2: Contains eight elements, from lithium (Li) to neon (Ne), showcasing a transition from metals to nonmetals.

Period 3: Includes elements from sodium (Na) to argon (Ar), reflecting similar trends as period 2.

B. Groups

The periodic table contains 18 groups, with elements in the same group exhibiting similar chemical behavior due to their valence electron configurations. Key groups include:

Group 1 (Alkali Metals): Highly reactive metals, such as lithium (Li) and sodium (Na).

Group 2 (Alkaline Earth Metals): Moderately reactive metals, including magnesium (Mg) and calcium (Ca).

Group 17 (Halogens): Nonmetals like fluorine (F) and chlorine (Cl) that readily form salts with metals.

Group 18 (Noble Gases): Inert gases like helium (He) and neon (Ne), known for their lack of reactivity.

3. Periodic Trends

The periodic table allows for the identification of several important trends in elemental properties, which arise from the structure of the table:

A. Atomic Radius

The atomic radius tends to decrease across a period from left to right due to increasing nuclear charge, which pulls electrons closer to the nucleus. Conversely, the atomic radius increases down a group as additional electron shells are added.

B. Ionization Energy

Ionization energy is the energy required to remove an electron from an atom. It generally increases across a period and decreases down a group. Elements with higher ionization energies are less likely to lose electrons and are typically nonmetals.

C. Electronegativity

Electronegativity measures an atom's ability to attract electrons in a chemical bond. It generally increases across a period and decreases down a group. Nonmetals, especially halogens, tend to have higher electronegativities than metals.

4. Applications of the Periodic Table

The periodic table is not just a mere classification tool; it has profound implications across various fields of science and technology:

A. Predictive Power

The periodic table enables chemists to predict the properties of unknown elements and compounds based on their position in the table. For example, the behavior of elements in a reaction can be anticipated based on their group placement.

B. Understanding Chemical Behavior

By studying periodic trends, scientists can understand how different elements interact and the types of compounds they form. This knowledge is crucial for fields such as materials science, pharmacology, and environmental chemistry.

C. Industrial Applications

In industry, the periodic table informs the selection of materials and chemical processes. For instance, the properties of metals, ceramics, and polymers can be predicted based on their positions in the table, guiding the development of new materials with desired characteristics.

5. Future Developments

As our understanding of atomic structure and chemical properties advances, the periodic table may undergo revisions and expansions. For example, the discovery of new synthetic elements or isotopes may lead to changes in the table's layout. Additionally, research into quantum mechanics and the behavior of elements at extreme conditions continues to enrich our understanding of chemical behavior.

The periodic table of elements is more than a reference tool; it is a comprehensive framework that encapsulates the relationships and trends among all known elements. From its historical roots to its modern applications, the periodic table continues to be a fundamental resource in chemistry and related sciences. By understanding the organization and significance of the periodic table, students and professionals can unlock the mysteries of chemical interactions, paving the way for future discoveries and innovations in science.

As you conclude your journey through the fundamental concepts of chemistry, remember the importance of the periodic table as both a map of the elements and a guide for navigating the intricate landscape of chemical behavior.

Conversion Tables and Common Constants

In the realm of chemistry, understanding measurements and the relationships between different units is crucial for accurate experimentation and effective communication of scientific data. This chapter provides an overview of conversion tables and common constants that serve as essential tools for chemists. These resources help in the conversion of units, ensuring consistency and precision in scientific calculations.

1. Importance of Conversion Tables

Conversion tables are charts that allow scientists and students to convert one unit of measurement to another, facilitating the understanding and application of scientific principles. These tables are particularly important in chemistry due to the diverse range of units used to describe quantities such as mass, volume, concentration, and energy. By using conversion tables, chemists can:

A. Facilitate Calculations

Many chemical reactions and processes are described using different units, such as grams, liters, or moles. Conversion tables help chemists easily switch between these units, ensuring accurate calculations in stoichiometry, concentration, and other areas.

B. Standardize Measurements

In scientific research and industry, standardized measurements are vital for consistency. Conversion tables help in converting local or regional units into universally accepted SI (International System of Units) units, promoting clarity in data presentation and interpretation.

C. Enhance Communication

In a global scientific community, chemists often collaborate across borders, using different measurement systems. Conversion tables enhance communication by providing a common reference point, allowing researchers to share data without confusion.

2. Common Conversion Tables

Several key conversion tables are widely used in chemistry. Here are some essential conversions that every chemist should be familiar with:

A. Mass Conversions

Grams to Moles: To convert grams to moles, divide the mass of the substance by its molar mass (grams/mole).

Example: For water (H_2O), the molar mass is approximately 18 g/mol. Therefore, 36 grams of water is equal to 2 moles (36 g ÷ 18 g/mol = 2 moles).

Kilograms to Grams: Multiply the mass in kilograms by 1,000.

Example: 2 kg = 2,000 g.

B. Volume Conversions

Milliliters to Liters: To convert milliliters to liters, divide the volume by 1,000.

Example: 500 mL = 0.5 L.

Cubic Centimeters to Liters: Since 1,000 cubic centimeters (cm^3) is equivalent to 1 liter (L), simply divide by 1,000 to convert.

Example: 250 cm^3 = 0.25 L.

C. Concentration Conversions

Molarity (M) to Mass Concentration (g/L): To convert from molarity to mass concentration, multiply the molarity by the molar mass.

Example: A 2 M solution of NaCl (molar mass = 58.44 g/mol) has a mass concentration of 116.88 g/L (2 M × 58.44 g/mol).

Parts per Million (ppm) to Concentration: To convert ppm to molarity, use the formula:

$$\text{ppm} = \left(\frac{\text{mass of solute (g)}}{\text{volume of solution (L)}} \right) \times 10^6$$

D. Energy Conversions

Caloric Energy to Joules: To convert from calories to joules, multiply by 4.184.

Example: 100 calories = 418.4 joules (100 × 4.184 = 418.4 J).

Kilojoules to Calories: To convert from kilojoules to calories, divide by 4.184.

Example: 1 kJ = 239.01 calories (1,000 J ÷ 4.184 = 239.01 cal).

3. Common Constants in Chemistry

In addition to conversion tables, several constants are fundamental in chemical calculations and must be well understood by chemists. These constants often serve as reference points in equations and conversions:

A. Avogadro's Number (N□)

Avogadro's number, approximately 6.022×10^{23}, is the number of atoms, molecules, or particles in one mole of a substance. This constant is pivotal in stoichiometric calculations, allowing chemists to relate macroscopic quantities (grams, liters) to microscopic amounts (moles, particles).

B. Gas Constant (R)

The ideal gas constant R is used in the ideal gas law equation $PV = nRT$ and is approximately:

$0.0821\,\mathrm{L \cdot atm/K \cdot mol}$

$8.314\,\mathrm{J/K \cdot mol}$ Depending on the units employed in the gas law calculations.

C. Molar Mass of Common Elements

Knowing the molar masses of common elements (e.g., carbon (C) = 12.01 g/mol, oxygen (O) = 16.00 g/mol, hydrogen (H) = 1.008 g/mol) is essential for stoichiometry and conversions involving mass and moles.

D. Specific Heat Capacities

The specific heat capacity of water is $4.18\,\mathrm{J/g \cdot {}^\circ C}$, a crucial value for calculations involving heat transfer, calorimetry, and thermodynamics.

4. Practical Applications

The knowledge of conversion tables and constants is vital in various practical applications:

A. Laboratory Measurements

In the laboratory, chemists routinely use conversion tables and constants to ensure accurate measurements, prepare solutions, and analyze reaction yields. For example, when diluting solutions, understanding how to convert molarity to mass concentration is essential for creating the desired concentrations.

B. Industrial Processes

In industrial settings, the conversion of raw materials into products requires precise calculations. Engineers and chemists use conversion tables to optimize processes, scale up reactions, and ensure product quality.

C. Research and Development

In research, accurate data collection and analysis are paramount. Conversion tables allow researchers to report findings in a consistent manner, facilitating peer review and collaboration across disciplines.

As this handbook concludes, the importance of conversion tables and common constants cannot be overstated. These tools are fundamental to the practice of chemistry, enabling precise calculations, promoting standardization, and enhancing communication within the scientific community. Whether in the classroom, laboratory, or industrial setting, mastery of conversion techniques and an understanding of key constants will empower students and professionals alike to navigate the complexities of chemical science with confidence and accuracy.

This foundational knowledge will serve as a stepping stone for further exploration in chemistry and its myriad applications, fostering a deeper appreciation for the scientific principles that govern the material world. As you move forward, remember that these tools will remain essential in your journey through the diverse and ever-evolving field of chemistry.

Glossary of Key Terms

In the study of chemistry, terminology is fundamental to understanding concepts, principles, and the language of science. A glossary of key terms serves as a valuable resource for students and practitioners, providing clear definitions and context for the various terms encountered throughout the subject. This glossary not only aids in learning but also enhances communication and comprehension in the field of chemistry. Below is an extensive glossary of essential chemistry terms.

1. Atom

The smallest unit of matter that retains the properties of an element. Atoms consist of a nucleus made of protons and neutrons, surrounded by electrons in orbitals.

2. Molecule

A group of two or more atoms bonded together by chemical forces. Molecules can be simple, like H_2 (hydrogen), or complex, like proteins.

3. Element

A pure substance that cannot be broken down into simpler substances by chemical means. Elements are defined by the number of protons in their atoms, known as the atomic number. Examples include hydrogen (H), carbon (C), and oxygen (O).

4. Compound

A substance formed when two or more elements chemically bond together in fixed proportions. Water (H_2O) and sodium chloride (NaCl) are common examples of compounds.

5. Ionic Bond

A type of chemical bond formed through the electrostatic attraction between oppositely charged ions. This typically occurs when an electron is transferred from one atom to another, as seen in sodium chloride (NaCl).

6. Covalent Bond

A chemical bond formed when two atoms share one or more pairs of electrons. This bond is common in molecules like water (H_2O) and carbon dioxide (CO_2).

7. Chemical Reaction

A process in which one or more substances (reactants) are transformed into one or more different substances (products) through the breaking and forming of chemical bonds.

8. Catalyst

A substance that increases the rate of a chemical reaction without being consumed in the process. Catalysts work by lowering the activation energy needed for the reaction to occur.

9. Stoichiometry

The branch of chemistry that deals with the calculation of reactants and products in chemical reactions based on the conservation of mass and the mole concept.

10. Mole

A unit of measurement in chemistry representing 6.022×10^{23} particles (atoms, molecules, ions, etc.). It is used to relate the macroscopic scale of substances to the atomic scale.

11. Concentration

The amount of a substance (solute) present in a given volume of solution. Common units of concentration include molarity (M), which is moles of solute per liter of solution, and parts per million (ppm).

12. pH

A measure of the acidity or basicity of a solution, ranging from 0 to 14. A pH of 7 is neutral, below 7 is acidic, and above 7 is basic (alkaline). pH is calculated using the negative logarithm of the hydrogen ion concentration.

13. Thermodynamics

The branch of physical chemistry that deals with the relationships between heat, work, temperature, and energy. It includes the laws of thermodynamics, which describe how energy is transferred and transformed.

14. Enthalpy

A measure of the total heat content of a system at constant pressure. It is used to determine the heat absorbed or released during chemical reactions.

15. Entropy

A measure of the disorder or randomness of a system. In thermodynamics, higher entropy indicates a greater degree of disorder and is associated with the second law of thermodynamics, which states that the total entropy of an isolated system always increases.

16. Periodic Table

A tabular arrangement of the chemical elements, organized by increasing atomic number, which reveals periodic trends in elemental properties. Elements in the same group share similar chemical characteristics.

17. Isomerism

The phenomenon where two or more compounds have the same molecular formula but different structural arrangements. Isomerism can be classified into structural isomerism and stereoisomerism.

18. Hydrocarbons

Organic compounds composed solely of hydrogen and carbon atoms. They can be categorized into aliphatic (straight-chain or branched) and aromatic compounds.

19. Functional Groups

Specific groups of atoms within molecules that determine the chemical reactivity and properties of those molecules. Common functional groups include hydroxyl (–OH), carbonyl (C=O), and carboxyl (–COOH).

20. Chemical Equilibrium

A state in a reversible chemical reaction where the rate of the forward reaction equals the rate of the reverse reaction, resulting in constant concentrations of reactants and products over time.

21. Solvent

The substance that dissolves a solute, resulting in a solution. Water is known as the "universal solvent" due to its ability to dissolve many substances.

22. Solution

A homogeneous mixture composed of two or more substances, where a solute is dissolved in a solvent. Solutions can be solid, liquid, or gas.

23. Colligative Properties

Properties of solutions that depend on the number of solute particles present, rather than the identity of the solute. These properties include boiling point elevation, freezing point depression, vapor pressure lowering, and osmotic pressure.

24. Bioinorganic Chemistry

A field of chemistry that studies the role of metals in biology, including the functions of metalloproteins, metalloenzymes, and metal ions in living organisms.

25. Nanotechnology

The manipulation and application of materials on the nanoscale (typically 1 to 100 nanometers). In chemistry, nanotechnology plays a role in developing new materials, catalysts, and drug delivery systems.

26. Quantum Chemistry

The branch of chemistry that applies the principles of quantum mechanics to explain the behavior of atoms and molecules. Quantum chemistry helps in understanding electronic structures, bonding, and reactivity.

27. Spectroscopy

A technique used to measure the interaction of light with matter, providing information about the structure, composition, and properties of substances. Common types include infrared (IR) spectroscopy, nuclear magnetic resonance (NMR) spectroscopy, and mass spectrometry.

28. Electrochemistry

The study of chemical processes that involve the movement of electrons, particularly in relation to redox reactions, electrolysis, and galvanic cells.

29. Safety Data Sheets (SDS)

Documents that provide essential information about hazardous substances, including handling, storage, and emergency measures. SDS are vital for ensuring safety in laboratories and industrial settings.

30. Laboratory Techniques

Methods and practices used in the laboratory to conduct experiments and analyses, including measuring, mixing, heating, and observing chemical reactions.

Made in United States
North Haven, CT
10 July 2025